Standardized Human Olfactory Thresholds

Standardized Human Olfactory Thresholds

Edited by

M. Devos, F. Patte, J. Rouault, P. Laffort

Laboratoire de Physiologie de la Chimioréception,
CNRS 91190 Gif-sur-Yvette (France)

and

L. J. Van Gemert

TNO-CIVO Food Analysis Institute, P.O. Box 360
3700 AJ Zeist (The Netherlands)

IRL PRESS
at
OXFORD UNIVERSITY PRESS
Oxford New York Tokyo

This book has been printed digitally and produced in a standard specification in order to ensure its continuing availability

OXFORD
UNIVERSITY PRESS

Great Clarendon Street, Oxford OX2 6DP

Oxford University Press is a department of the University of Oxford.
It furthers the University's objective of excellence in research, scholarship,
and education by publishing worldwide in

Oxford New York

Auckland Cape Town Dar es Salaam Hong Kong Karachi
Kuala Lumpur Madrid Melbourne Mexico City Nairobi
New Delhi Shanghai Taipei Toronto

With offices in

Argentina Austria Brazil Chile Czech Republic France Greece
Guatemala Hungary Italy Japan South Korea Poland Portugal
Singapore Switzerland Thailand Turkey Ukraine Vietnam

Published in the United States
by Oxford University Press Inc., New York

Reprinted 2011

ISBN 978-0-19-963146-9

Printed and bound in Great Britain by CPI Antony Rowe,
Chippenham and Eastbourne

Contents

Compilation and standardization method

Introduction

Two important compilations of human olfactory thresholds have recently been published, one by Fazzalari (1978) and one by Van Gemert and Nettenbreijer (1977) and Van Gemert (1984). These collections are useful sources of original data.

However, a problem encountered with this type of compilation is that different authors obtain widely different values for the same substance. This finding, *a priori* disappointing, was optimized by the observation of Laffort (1963), who showed that these differences could be considerably reduced by assigning a specific weighting coefficient to each author's results. Using this method, Laffort proposed a collection of standardized thresholds for about 200 substances. The same principle was used by Patte *et al.* (1975) to obtain standardized exponents of the power law for 110 substances. The object of the present study was to provide an inventory of weighted and averaged human olfactory threshold which is as exhaustive and homogeneous as possible.

Methods

Sources of data

Values considered to be original are those reported by Van Gemert and Nettenbreijer (1977) and by Van Gemert (1984), restricted to the olfactory thresholds in the gas phase (i.e. leaving aside the thresholds in the aqueous phase). This original compilation and its supplement contain a total of 2075 odour threshold values in air of 641 chemical compounds, gathered from 372 references. In some rare cases, we anticipated the corrections or complements to be published in the next supplement. When the original data were presented as a range of values, we calculated the geometrical mean of the extreme concentrations. When the detection and recognition thresholds were given, we retained only the detection threshold.

Log_{10} units were used, as they are easier to handle than classical concentration units. Results were expressed using two types of log units, both having in common the denomination of p.ol, the abbreviation of 'olfactory power', as proposed by Laffort (1963). The p.ol(volume) is the negative log of the concentration expressed in volumic (or molar) fractions. It may be directly transposed into currently used units such as ppm(volume) and ppb(volume): thresholds of 1 ppm and 1 ppb are equal to p.ol = 6 and p.ol = 9 respectively. The limit of olfactory effectiveness (odourlessness) at normal pressure (1 litre per litre), is equivalent to p.ol(volume) = 0. The p.ol(mass) is equal to the negative log of the concentration expressed in g/l. An olfactory threshold of 1 mg/m^3 is equivalent to p.ol(mass) = 6.

Both units are distinct from those proposed by Laffort (1963) under the name of p.ol (without specification), which was based on the method used to express pH: the negative log of the molar concentration. This definition is more rigorous (it is valid irrespective of the atmospheric pressure); however, it is less practical (no simple correspondence with the ppm, ppb or mg/m^3, and odourlessness does not correspond to the value 0).

Calculation of the standard thresholds

Standard threshold were estimated by a stepwise process. The general method consists of successive constitution of enlarged pools of homogeneous data and comparison of the data with those of references not yet incorporated.

Each reference was characterized by a weighting coefficient Δ which is a measure of the systematic deviation from the average pool. Proximity between data was provided by a reference (weighted with Δ) and the average pool was estimated by an index of profile similarity PS with values between -1 and $+1$. The stepwise process and the statistics Δ and PS are described in Appendix A.

Results

Starting with the data of Leonardos *et al.* (1969), who used the natural sniffing technique with a relatively large number of substances, ten steps were necessary to obtain the final inventory which includes the averaged, standardized, olfactory thresholds of 529 substances, determined from data from 120 references (out of a total of 641 substances in 372 references). These 120 references can be reduced to 105 distinct sources or references. We limited the number of steps to 10, as the gain of references and substances tended towards zero.

Table 1.1 shows the list of the weighting coefficients Δ obtained for the 105 selected references. The values N and N′ are also reported in Table 1.1 for each reference. N is the number of substances in common with at least one of the others selected, and N′ is the total number of substances studied in that reference. Table 1.1 also includes the PS values, obtained on the basis of N for each reference compared to the final pool. Substances studied by only one author, the results of which artificially increased this average similarity of profile,

Table 1.1 Weighting coefficients Δ, arranged in alphabetical and chronological order according to the first author. N represents the overlapping of substances studied in each reference, with the pool of substances studied by at least one of the other selected references. N′ represents the total number of substances studied in each selected reference. The profile similarity indexes PS are based on N.

	Authors	Δ	N	N′	PS
1	Allison and Katz (1919)	2.58	20	23	0.90
2	Amoore (1977)	−0.76	3	5	0.95
3	Anon (1980)	−0.97	46	48	0.92
4	Anrooij, van (1931)	−0.74	11	12	0.89
5	Appell (1969)	−2.39	19	37	0.89
6	Backman (1917)	0.00	113	185	0.92
7	Backman (1918)	−0.48	5	5	0.98
8	Baikov (1963)	−0.16	2	2	1.00
9	Baldus (1936)	−1.10	13	15	0.87
10	Basmadshijewa *et al.* (1968, 1970)	−0.85	3	3	0.95
11	Beck (1959)	−0.27	4	4	0.94
12	Bedborough and Trott (1979)	−0.70	10	11	0.94
13	Blanchard (1976)	0.77	7	7	0.96
14	Blinova (1965)	0.57	3	4	0.99
15	Bocca and Battiston (1964)	0.30	2	3	0.99
16	Buchberg *et al.* (1961)	0.12	3	3	0.98
17	Cain (1969)	1.55	4	4	0.95
18	Corbit and Engen (1971)	0.96	3	3	0.99

	Authors	Δ	N	N′	PS
19	Cormack *et al.* (1974)	−0.37	4	4	0.95
20	Deadman and Prigg (1959)	−0.32	14	23	0.93
21	Dravnieks and O'Donnell (1971)	0.47	4	4	0.95
22	Dravnieks and Laffort (1972)	0.10	30	31	0.91
23	Dravnieks (1974)	1.20	28	29	0.95
24	Elfinova (1966)	−0.71	2	2	1.00
25	England *et al.* (1978)	1.05	2	4	1.00
26	Eriksson *et al.* (1976)	0.29	4	4	0.96
27	Fieldner *et al.* (1921)	0.95	3	3	0.99
28	Gavaudan and Poussel (1966)	−0.46	4	4	0.96
29	Geier (1936)	−0.42	6	8	0.91
30	Gofmekler (1960)	−0.55	5	5	0.89
31	Goldenberg (1967)	−1.23	5	5	0.95
32	Grijns (1919)	1.24	11	11	0.94
33	Guadagni (1966)	−0.24	9	9	0.85
34	Gusev (1965)	−0.39	3	3	0.97
35	Hall and Andersson (1983)	0.47	22	28	0.95
36	Hellman and Small (1973, 1974)	−0.52	50	93	0.94
37	Hendriks (1979)	−0.33	8	8	0.94
38	Henning (1924)	−0.81	9	9	0.87
39	Hermanides (1909)	−0.55	8	8	0.95
40	Hesse (1926)	1.57	6	6	0.88
41	Hesse (1927)	1.62	4	4	0.91
42	Huijer (1918)	1.06	5	5	0.96
43	Janicek *et al.* (1960)	1.58	11	12	0.93
44	Jones (1954)	1.68	2	2	0.99
45	Jones (1955*a*)	0.15	3	3	0.98
46	Jones (1955*b*)	1.51	20	20	0.94
47	Jung (1936)	−0.47	18	19	0.92
48	Katz and Talbert (1930)	0.20	34	54	0.90
49	Kendall *et al.* (1968)	−0.41	8	12	0.85
50	Kerka and Humphreys (1956)	−0.38	4	4	0.94
51	Kinkead *et al.* (1971*a,b*)	−0.93	2	2	0.99
52	Kniebes *et al.* (1969)	−0.20	5	5	0.97
53	Knuth (1973)	−0.23	7	7	0.96
54	Köster (1971)	1.50	15	19	0.84
55	Laffort and Dravnieks (1973)	0.41	21	24	0.96
56	Laing (1975)	1.38	5	5	0.96
57	Laing *et al.* (1978)	1.12	7	8	0.95
58	Laing (1982)	0.65	3	3	0.97
59	Lehman and Schmidt–Kehl (1936)	1.15	5	6	0.97
60	Leonardos *et al.* (1969)	−0.01	46	53	0.93
61	Logtenberg (1978)	−0.53	4	5	0.93
62	Makeicheva (1978)	−1.10	4	4	0.95

	Authors	Δ	N	N′	PS
63	May (1966)	1.19	33	34	0.95
64	Mitsumoto (1926)	1.04	6	6	0.91
65	Morimura (1934)	1.17	5	5	0.95
66	Moschandreas and Jones (1983)	0.05	4	5	0.96
67	Mullins (1955)	1.75	19	24	0.93
68	Nader (1958)	−0.25	4	4	0.91
69	Naus (1962)	−0.62	11	11	0.91
70	Naus (1982)	−0.48	21	21	0.89
71	Nevers and Oister (1965)	0.00	3	3	0.98
72	Nishida *et al.* (1975)	0.15	4	4	0.95
73	Ohma (1922)	0.18	6	6	0.94
74	Passy (1892*a*,*b*,*c*)	0.15	18	19	0.90
75	Passy (1893*a*,*b*,*c*)	0.03	14	14	0.92
76	Patty and Yant (1929)	1.10	5	5	0.97
77	Patty (1962*a*,*b*)	−0.04	2	3	1.00
78	Pliska (1962)	1.98	5	7	0.92
79	Pliska and Janicek (1965)	2.80	17	17	0.90
80	Punter (1983)	0.40	42	58	0.92
81	Randebrock (1971)	−1.85	9	12	0.88
82	Rocen (1920)	0.55	5	5	0.91
83	Sales (1958)	0.06	3	3	0.99
84	Scherberger *et al.* (1958)	1.42	12	12	0.95
85	Schley (1934)	−0.37	7	10	0.95
86	Schwarz (1955)	0.06	6	8	0.96
87	Shalamberidze (1967)	−0.09	2	2	0.99
88	Smith and Hochstettler (1969)	−0.44	3	3	0.96
89	Smolczyk and Cobler (1930)	0.21	3	4	0.95
90	Stalker (1963)	−0.92	3	3	0.99
91	Stephens (1971)	−0.09	4	5	0.95
92	Stone (1963*a*,*b*,*c*)	0.78	5	5	0.97
93	Stone and Bosley (1965)	0.95	2	2	0.99
94	Stone *et al.* (1967)	−0.27	6	7	0.94
95	Stone *et al.* (1972)	0.08	6	8	0.93
96	Stuiver (1958)	−1.29	22	28	0.91
97	Takhirov (1969, 1974)	−0.81	7	7	0.91
98	Tempelaar (1913)	0.59	24	33	0.88
99	Teranishi *et al.* (1974)	−0.09	22	29	0.94
100	Thomas *et al.* (1943)	0.10	3	3	0.98
101	Tkachev (1963, 1969, 1970, 1978)	−0.75	5	5	0.97
102	Valentin (1848, 1850)	1.14	4	4	0.94
103	Wilby (1969)	−0.08	14	18	0.94
104	Wolf *et al.* (1956)	2.20	2	2	1.00
105	Zwaardemaker (1914)	−0.08	19	19	0.86

were therefore excluded from this calculation. The significances of the selected data can be checked using Table A.1 in Appendix A. It should be noted that only 57 references were left out of the pool because of lack of consistency (p of $PS > 0.01$). The 195 remaining references were not taken into account because of insufficient cross checking of the substances they included with those of the pool (less than 2).

The original data and the weighted and averaged data are reported in Appendix B. The same nomenclature and the same order of presentation are adopted as used by Van Gemert and Nettenbreijer (1977) and by Van Gemert (1984). We have added the *Handbook of chemistry and physics* (50th edn) reference number and the molecular weight M. Thus p.ol(mass) can easily be calculated from p.ol(volume) using the equation:

$$\text{p.ol(mass)} = \text{p.ol(volume)} + 1.38 - \log(M) \qquad (1)$$

(The constant 1.38 is equal to $\log_{10}$ of the molar volume of the perfect gases at 20°C.)

The distribution of Δ values is given by the histogram in Fig. 1.1, and the distribution of the averaged weighted thresholds by the histogram in Fig. 1.2.

Discussion

Methods

The principle aim of the PS index is to make comparison possible between authors who studied only two substances in common. Such comparisons cannot be made with classical tests; for example, the Spearman correlation coefficient requires at least three elements of comparison. This PS index has been already used in a French document concerning olfactory air pollution (AFNOR 1986). Table A.1 in Appendix A will be completed and refined in a second, more specialized article, which will include a more general expression of the PS concept, valid under all circumstances, and not only for olfactory thresholds (i.e. without the hypothesis concerning the range values).

Results

The scattering of the observed weighting coefficients may be

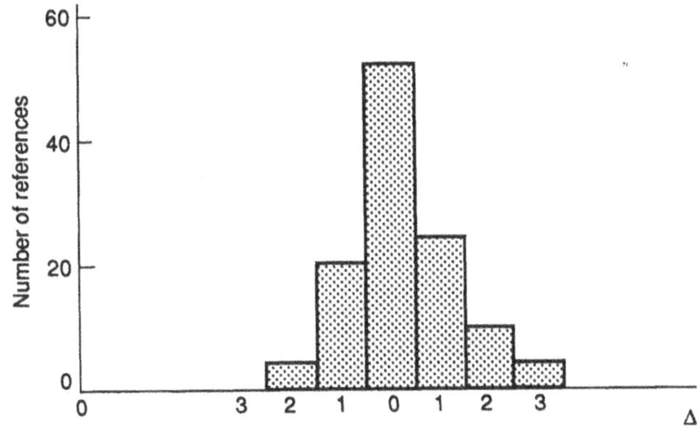

Fig. 1.1 Distribution of weighting coefficients Δ for the 105 selected references.

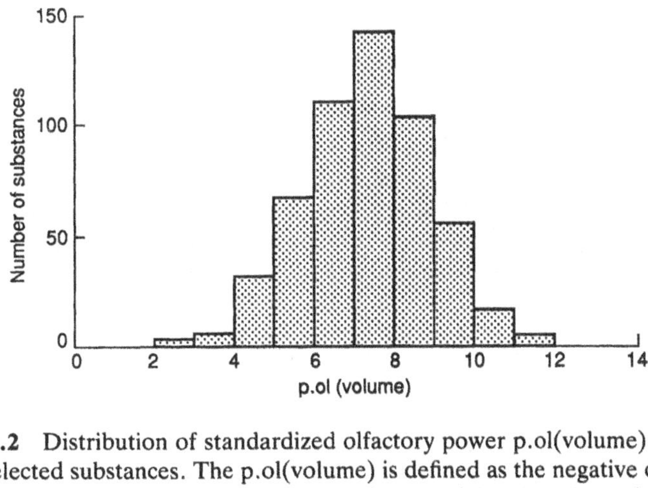

Fig. 1.2 Distribution of standardized olfactory power p.ol(volume) for the 529 selected substances. The p.ol(volume) is defined as the negative decimal log of concentration expressed in volumic (or molar) fraction. A p.ol(volume) of 0, 6, 9 corresponds to odourlessness at normal atmospheric pressure, and to thresholds of 1 ppm and 1 ppb respectively.

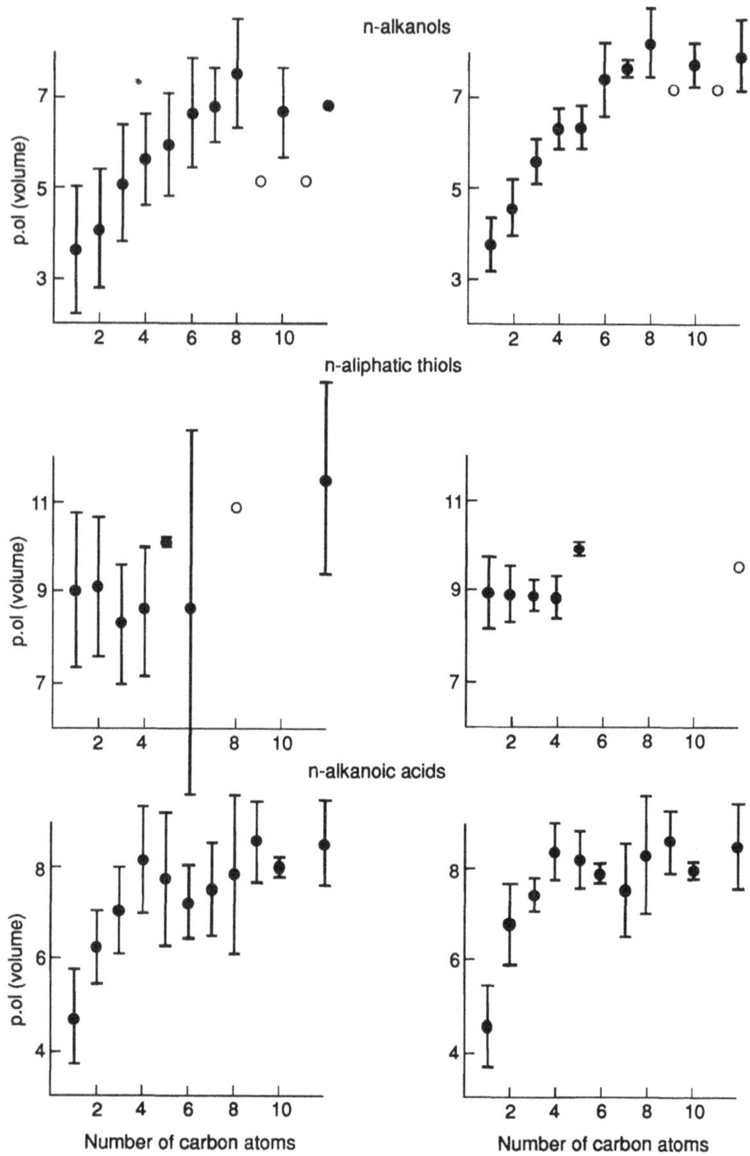

Fig. 1.3 Comparison of the olfactory power p.ol(volume) for the substances of three homologous series before and after using weighting coefficients. Full circles correspond to substances reported in at least two references. Open circles correspond to substances reported in one reference. Confidence intervals are given by $\pm\sigma$ (standard deviation).

explained in many ways. For instance, systematic errors in evaluating the dilutions will generate threshold values which are too low or too high. With static successive dilution techniques, the trend will be positive (May 1966), because of adsorption phenomena. Subject-dependent factors affecting the responses include the well-known effect of female sex hormone impregnation, and the satiety, wide-awake and peripheral vasodilatation states. Among the main factors which condition the olfactory stimuli, known to have an influence on responses, are the temperature and the flow rate. For example, with low flow rates, the trend of Δ will be positive (Mullins 1955) in comparison with the thresholds obtained with natural sniffing, which were chosen as a reference. It is also known that the presentation procedure (forced choice or not), modifies the level of the values obtained. Most of the factors modulating olfactory responses have been reviewed by Laffort (1966) and were taken into account in AFNOR (1986). A theoretical justification of the weighting coefficients Δ reported in Table 1.1 could possibly be undertaken for each reference; this was not the purpose of the present work, which was limited to a purely experimental approach.

From a strictly practical point of view, the increase in reliability of these data averaged after weighting, in addition to the PS values in Table 1.1, is visualized in Fig. 1.3. We took the thresholds of three homologous series whose confidence intervals before treatment were particularly disturbing, as shown by Van Gemert and Nettenbreijer (1977). There was a considerable reduction in these confidence intervals after standardization.

Conclusion

Weighting by selected reference, and averaging of the threshold values gave an inventory that was more homogeneous, though a little less exhaustive, than that of Van Gemert and Nettenbreijer (1977) and Van Gemert (1984).

References

Afnor (1986). *Qualité de l'air. Méthode de mesurage de l'odeur d'un effluent gazeux. Détermination du facteur de dilution au seuil de perception.* Norme X 43–101, AFNOR, Tour Europe Cedex 7, 92080 Paris la Défense, France.

Allison, V. C. and Katz, S. H. (1919). An investigation of stenches and odours for industrial purposes. *J. Ind. Eng. Chem.* **11,** 336–8.

Amoore, J. E. (1977). Specific anosmia and the concept of primary odors. *Chem. Senses Flavour*, **2,** 267–81.

Anrooij, van: see Van Anrooij.

Anon (1980). *Reports of studies on the measurement of the offensive odours (from 1972–1980)* (in Japanese). Japan Environment Agency, Tokyo.

Appell, L. (1969). Physical foundation in perfumery VIII. The minimum perceptible. *Am. Perfum. Cosmet.* **84,** (3) 45–50.

Backman, E. L. (1917). Experimentella undersökningar öfver luktsinnets fysiologi. *Upsala Laekarefoeren. Foerh.* **22,** 319–470.

Backman, E. L. (1918). The olfactology of the methylbenzol series. *Onderzoekingen gedaan in het Physiologisch Laboratorium der Utrechtse Hogeschool, S5,* **18,** 349–64.

Baikov, B. K. (1963). Experimental data for substantiating the maximum permissible concentration of carbon disulphide in combination with hydrogen disulphide in the atmospheric air (in Russian). *Gig. Sanit.* **28**(3), 3–8.

Baldus, C. (1936). *Untersuchung über Geruchsswellen.* Thesis. University of Würzburg.

Basmadshijewa, K., Argirowa, M. and Dawidkswa, E. (1970). Hygienische Begründung des MIK-Wertes von Methylendichlorid in der atmosphärischen Luft. *Wiss. Z. Humboldt-Univ. Berlin, Math. Naturwiss. Reihe,* **19,** 469–70.

Basmadzhieva, K. and Argirova, M. (1968). Maximum permissible concentration of hydrogen sulfide and phenol during their simultaneous presence in the atmosphere. *Khig. Zdraveopazvane*, **11,** 237–44.

Beck, H. (1959). *Experimentelle Ermittlung von Geruchsschwellen einiger wichtiger Reizgase (Chlor, Schwefeldioxyd, Ozon, Nitrose) und Erscheinungen bei Einwirkung geringer Konzentrationen auf den Menschen.* Inaugural-Dissertation. University of Würzburg.

Bedborough, D. R. and Trott, P. E. (1979). *The sensory measurement of odours by dynamic dilution.* Report No: LR 299 (AP), Warren Spring Laboratory, Stevenage.

Blanchard, J. (1976). Relazioni tra l'odore di un gas ed il suo tenore in mercaptani. *Gas (Rome)*, **26,** 287–309.

Blinova, E. (1965). Industrial standards for substances emitting strong odors. *Hyg. Sanit. USSR*, **30**(1–3), 18–22.

Bocca, E. and Battiston, M. N. (1964). Odour perception and environment conditions. *Acta Oto-Laryngol*, **57,** 391–400.

Buchberg, H., Jones, M. H., Lindh, K. G. and Wilson, K. W. (1961). *Air pollution studies with simulated atmospheres.* Report no 61–44, California University, Los Angeles.

Cain, W. S. (1969). Odor intensity differences in the exponent of the psychophysical function. *Percept. Psychophys.* **6,** 6A, 349–54.

Corbit, T. E. and Engen, T. (1971). Facilitation of olfactory detection. *Percept. Psychophys.* **10,** 433–6.

Cormack, D., Dorling, T. A. and Lynch B. W. J. (1974). Comparison of techniques for organoleptic odour-intensity assessment. *Chem. Ind. (London)*, **21,** 857–61.

Deadman, K. A. and Prigg, J. A. (1959). *The odour and odorization of gas.* Research Communication GC59, The Gas Council, London.

Dravnieks, A. (1974). A building-block model for the characterization of odorant molecules and their odors *Ann. NY Acad. Sci.* **237,** 144–63.

Dravnieks, A. and Laffort, P. (1972). Physio-chemical basis of quantitative and qualitative odor discrimination in humans. In *Olfaction and taste IV* (ed. D. Schneider), pp.142–8. Proc. Fourth Internat. Symp. August 1971, Wissenschaftliche Verlagsgesellschaft MBH, Stuttgart. (Converted data taken from: Patte, F., Etcheto, M. and Laffort, P. (1975). Selected and standardized values of suprathreshold odour intensities for 110 substances. *Chem. Senses Flavour*, **1,** 238–305.)

Dravnieks, A. and O'Donnell, A. (1971). Principles and some techniques of high resolution headspace analysis. *J. Agric. Food Chem.* **19,** 1049–56.

Elfimova, E. V. (1966). Biological activity of low atmospheric concentrations of isopropylbenzene and benzene. Vop. Gig. Planirovki San. Okhr. Atmos. Vozdukha, Moscow, 148–58. (From: *Chem. Abstr.* (1967) **66,** 93947.)

England, A., Laing, D. G. and Panhuber, H. (1978). *Odorous pollutants.* Report of Research 1977–1978, p.23, CSIRO Division of Food Research.

Eriksson, C. E., Lundgren, B. and Vallentin, K. (1976). Odor detectability of aldehydes and alcohols originating from lipid oxidation. *Chem. Senses Flavor*, **2,** 3–15. (Converted data taken from Hall and Andersson (1983)).

Fazzalari, F. A. (1978). *Compilation of odor and taste threshold values data.* ASTM Data series DS4 8A. Philadelphia, (USA).

Fieldner, A. C., Katz, S. H. and Kinney, S. P. (1921). *Gas masks for gases met in fighting fires.* Technical Paper No 248, US Bureau of Mines.

Gavaudan, P. and Poussel, H. (1966). Conditions physico-chimiques du déclenchement des sensations olfactives. *CR Congr. Natl. Soc. Savantes Sect. Sci.* **90,** 607–13.

Geier, F. (1936). *Beitrag zur Bestimmung von Geruchsschwellen.* Thesis. University of Würzburg.

Gofmekler, V. A. (1960). Data for substantiating the maximum permissible concentration of acetates in the atmosphere (in Russian). *Gig. Sanit.* **25** (4), 9–15.

Goldenberg, D. M. (1967). *Geruchswahrnehmung und Schwellen von Duftgemischen beim Menschen.* Hals-, Nasen- und Ohrenheilkunde, Zwanglose Schriftenreihe, Heft 19 Barth, Leipzig.

Grijns, G. (1919). Y-a-t-il une relation entre le pouvoir absorbant à l'égard de la chaleur rayonnante et le pouvoir odorant des substances? *Arch. Neerl. Physiol.* **3,** 377–90.

Guadagni, D. G. (1966). Private communication cited in Teranishi, R. (1967). High resolution gas chromatography in aroma research. *Am. Perfum. Cosmet.*, **82,** (9), 43–4, 46, 48, 51.

Gusev, I. S. (1965). Reflective effects of microconcentrations of benzene, toluene, xylene and their comparative assessment. *Hyg. Sanit. USSR*, **30**(10–12), 331–6.

Hall, G. and Andersson, J. (1983). Volatile fat oxidation products I. Determination of odour thresholds and odour intensity functions by dynamic olfactometry. *Lebsensm. Wiss. Technol.* **16,** 354–61.

Hellman, T. M. and Small, F. H. (1973). Characterization of petrochemical odors. *Chem. Eng. Progr.* **69,** (9), 75–7.

Hellman, T. M. and Small, F. H. (1974). Characterization of the odor properties of 101 petrochemicals using sensory methods. *J. Air Pollut. Contr. Assoc.* **24,** 979–82.

Hendriks, A. P. J. (1979). *Zelfadaptatie en kruisadaptatie met enige primaire en niet-primaire geurstoffen.* Unpublished report, Psychological Laboratory, University of Utrecht.

Henning, H. (1924). *Der Geruch (zweite Ausgabe)*. Barth, Leipzig.

Hermanides, J. (1909). *Over de constanten der in de olfactometrie gebruikelijke negen standaardgeuren.* Thesis. University of Utrecht.

Hesse, W. (1926). Bestimmung von Geruchsschwellen in absoluten Werten. *Z. Hals-, Nasen-, Ohrenheilk.*, **16,** 359–73.

Hesse, W. (1927). Ein einfacher Apparat zur Bestimmung von feruchs-schwellen in absoluten Werten. *Z. Hals-, Nasen-, Ohrenheilk.*, **19**, 348–52.

Huijer, H. (1918). *De olfactologie van aniline en homologen.* Thesis. University of Utrecht.

Janicek, G., Pliska, V. and Kubatova, J. (1960). Olfactometric estimation of the threshold of perception of odorous substances by a flow olfactometer (in Czech.). *Cesk. Hyg.*, **5,** 441–7.

Jones, F. N. (1954). An olfactometer permitting stimulus specification in molar terms. *Am. J. Psychol.* **67,** 147–51.

Jones, F. N. (1955*a*). The reliability of olfactory thresholds obtained by sniffing. *Am. J. Psychol.* **68,** 289–90.

Jones, F. N. (1955*b*). Olfactory absolute thresholds and their implications for the nature of the receptor process. *J. Psychol.* **40,** 223–7.

Jung, J. (1936). *Untersuchung über Geruchsschwellen.* Thesis. University of Würzburg.

Katz, S. H. and Talbert, E. J. (1930). *Intensities of odors and irritating effects of warning agents for inflammable and poisonous gases.* Technical Report No 480, US Bureau of Mines.

Kendall, D. A., Leonardos, G. and Rubin, E. R. (1968). *Parameters affecting the determination of odor thresholds.* Report C-68988, Arthur D. Little, Inc., Cambridge (USA).

Kerka, W. F. and Humphreys, C. M. (1956). Temperature and humidity effect on odor perception. *Heat. Piping Air Cond.* **28,** 129–36.

Kinkead, E. R., Pozzani, U. C., Geary, D. L. and Carpenter, C. P. (1971*a*). The mammalian toxicity of ethylidenenorbornene (5-ethylidenebicyclo (2,2,1) hept-2-ene). *Toxicol. Appl. Pharmacol.* **20,** 250–9.

Kinkead, E. R., Pozzani, U. C., Geary, D. L. and Carpenter, C. P. (1971*b*). The mammalian toxicity of dicyclopentadiene. *Toxicol. Appl. Pharmacol.* **20,** 552–61.

Kniebes, D. V., Chisholm, J. A. and Stubbs, R. C. (1969). Odorizzanti del gas naturale: fattori di sensibilita' olfattiva. *Gas (Rome)*, **19,** 211–17.

Knuth, H. W. (1973). *Personal communication.* Institut für Chemische Technologie, TU Clausthal.

Köster, E. P. (1971). *Adaptation and cross-adaptation in olfaction*. Thesis. University of Utrecht.

Laffort, P. (1963). Essai de standardisation des seuils olfactifs humains pour 192 corps purs. *Arch. Sci. Physiol.* **17,** 75–105.

Laffort, P. (1966). *Les facteurs de l'activité quantitative des molécules odorantes*. Thesis. University of Paris-Orsay, France.

Laffort, P. and Dravnieks, A. (1973). An approach to a physico-chemical model of olfactory stimulation in vertebrates by single compounds. *J. Theor. Biol.* **38,** 335–45.

Laing, D. G. (1975). A comparative study of the olfactory sensitivity of humans and rats. *Chem. Senses Flavour*, **1,** 257–69.

Laing, D. G. (1982). Characterisation of human behaviour during odour perception. *Perception*, **11,** 221–30.

Laing, D. G., Panhuber, H. and Baxter, R. I. (1978). Olfactory properties of amines and n-butanol. *Chem. Senses Flavour*, **3,** 149–66.

Lehmann, K. B. and Schmidt–Kehl, L. (1936). Die 13 wichtigsten Chlorkohlwasserstoffe der Fettreihe vom Standpunkt der Gewerbehygiene. *Arch. Hyg. Bakteriol.* **116,** 131–268.

Leonardos, G., Kendall, D. and Barnard, N. (1969). Odor threshold determinations of 53 odorant chemicals. *J. Air Pollut. Contr. Assoc.* **19**, 91–5.

Logtenberg, M.Th. (1978). Onderzoek naar de nauwkeurigheid van geurconcentratiemetingen met de TNO Snuifkar. Unpublished report, Division for Physical and Chemical Technology TNO, Apeldoorn.

Makeicheva, N. A. (1978). Data for hygienic standardization of a mixture of furfurol, formaldehyde, phenol and acetone in the atmosphere (in Russian). *Gig. Sanit.* **9,** 3–7.

May, J. (1966). Geruchsschwellen von Lösemitteln zur Bewertung von Lösemittelgerüchen in der Luft. *Staub*, **26,** 385–9.

Mitsumoto, T. (1926). Olfaktometrische Untersuchungen. *Z. Sinnesphysiol.* **57,** 114–165, 317.

Morimura, S. (1934). Untersuchung über den Geruchssin. *Tohoku J. Exp. Med.* **22,** 417–48.

Moschandreas, D. J. S. and Jones, D. R. (1983). Odour detection, recognition and fatigue. In *Proceedings of VIth world congress on air quality—II Atmospheric pollution and health*, Paris (France) SEPIC, 392–9.

Mullins, L. J. (1955). Olfaction. *Ann. NY Acad. Sci.* **62,** 247–76.

Nader, J. S. (1958). An odor evaluation apparatus for field and laboratory use. *J. Am. Ind. Hyg. Assoc.* **19,** 1–7.

Naus, A. (1962). In Naus, A. (1975) *Olfactoric properties of industrial matters*. Charles University, Prague.

Naus, A. (1982). Olfactory thresholds of industrial substances (in Czech.). *Prac. Lek.* **34,** 217–8.

Nevers, A. D. and Oister, W. H. (1965). Problems in the critical comparison of odor intensities. *Proc. Oper. Sect. Am. Gas. Assoc.*

Nishida, K., Honda, T. and Bota, K. (1975). *New apparatus on the measurement of odor intensity*. Paper 75-55.3, 68th Ann. Meet. Air Pollut. Control Assoc., June 15–20, Boston.

Ohma, S. (1922). La classification des odeurs aromatiques en sous-classes. *Arch. Neerl. Physiol.*, **6,** 567–90.

Passy, J. (1892*a*). Sur quelques minimums perceptibles d'odeurs. *CR Hebd. Séances Acad. Sci.* **114,** 786–8.

Passy, J. (1892*b*). Sur la perception des odeurs. *CR Soc. Biol.* **44,** 239–43.

Passy, J. (1892*c*). Les propriétés odorantes des alcools de la série grasse. *CR Hebd. Séances Acad. Sci.* **114,** 1140–3.

Passy, J. (1893*a*). Pouvoir odorant du chloroforme, du bromoforme et de l'iodoforme. *CR Hebd. Séances Acad. Sci.* **116,** 769–70.

Passy, J. (1893*b*). Forme périodique du pouvoir odorant dans la série grasse. *CR Hebd. Séances Acad. Sci.* **116,** 1007–10.

Passy, J. (1893*c*). L'odeur dans la série grasse. *CR Soc. Biol.* **45,** 479–81.

Patte, F., Etcheto, M. and Laffort, P. (1975). Selected and standardized values of suprathreshold odour intensities for 110 substances. *Chem. Senses Flavour*, **1,** 283–305.

Patty, F. A. (1962*a*). Alkaline materials. In *Industrial hygiene and toxicology* Volume II: *Toxicology* (ed. F. A. Patty), pp.859–69. Interscience, New York/London.

Patty, F. A. (1962*b*). Inorganic compounds of oxygen, nitrogen and carbon. In *Industrial hygiene and toxicology* Volume II: *Toxicology*, (ed. F. A. Patty), pp.911–40, Interscience, New York/London.

Patty, F. A. and Yant, W. P. (1929). *Odor intensity and symptoms produced by commercial propane, butane, pentane, hexane and heptane vapor*. R.I. 2979, US Bureau of Mines.

Pliska, V. (1962). Olfaktometrische Studien IV. Fehleranalyse olfaktometrischer Messungen. *Sb. Vys. Sk. Chem.-Technol. Praze, Potraviny*, **6,** (1), 37–46.

Pliska, V. and Janicek, G. (1965). Die Veränderungen der Wahrnehmungsschwellen-Konzentration der Riechstoffe in einiger homologischen Serien. *Arch. Int. Pharmacodyn. Ther.* **156,** 211–16.

Punter, P. H. (1983). Measurement of human olfactory thresholds for several groups of structurally related compounds. *Chem. Senses*, **7**, 215–35.

Randebrock, E. M. (1971). Molecular theory of odor with the α-helix as potential perceptor. In *Gustation and olfaction* (eds. G. Ohloff and A. F. Thomas) pp.111–25. Academic Press, London/New York.

Rocen, E. (1920). Contribution to the localisation of 'sweet smell'. *Scand. Arch. Physiol.* **40,** 129–44.

Rouault, J. and Capy, P. (1986). Comment évaluer un nombre de catégories par échantillonnage. *Annales d'économie et de statistique*, **4,** 111–24.

Sales, M. (1958). *Odeur et odorisation des gaz* (IGU/36-58). 7e Congrès International de l'Industrie du Gaz, Rome.

Scherberger, R. F., Happ, G. P., Miller, F. A. and Fassett, D. W. (1958). A dynamic apparatus for preparing air-vapor mixtures of known concentrations. *Am. Ind. Hyg. Ass. Quart.* **19,** 494–8.

Schley, O.-H. (1934). *Untersuchung über Geruchsschwellen.* Thesis. University of Würzburg.

Schwarz, R. (1955). Uber die Riechschärfe der Honigbiene. *Z. Vergl. Physiol.* **37,** 180–210.

Shalamberidze, O. P. (1967). Reflex action of a mixture of sulfur dioxide and nitrogen dioxide (in Russian). *Gig. Sanit.* **32,** (7), 9–13.

Smith, H. O. and Hochstettler, A. D. (1969). Determination of odor thresholds in air using ^{14}C-labeled compounds to monitor concentrations. *Environ. Sci. Technol.* **3,** 169–70.

Smolczyk, E. and Cobler, H. (1930). Chemischer Nachweis von Atemgiften und subjektive Empfindlichkeit. *Gasmaske*, **2,** 27–33.

Stalker, W. W. (1963). Defining the odor problem in a community. *J. Am. Ind. Hyg. Assoc.* **24,** 600–5.

Stephens, E. R. (1971). Identification of odors from cattle feed lots. *Calif. Agric.* **25,** (1), 10–11.

Stone, H. (1963*a*). Techniques for odor measurement: olfactometric vs. sniffing. *J. Food Sci.* **28,** 719–25.

Stone, H. (1963*b*). Determination of odor difference limens for three compounds. *J. Exp. Psychol.* **66**, 466–73.

Stone, H. (1963*c*). *Some factors affecting olfactory sensitivity and odor intensity.* Thesis. University of California.

Stone, H. and Bosley, J. J. (1965). Olfactory discrimination and Weber's law. *Percept. Mot. Skills*, **20,** 657–65.

Stone, H., Pryor, G. T. and Colwell, J. (1967). Olfactory detection thresholds in man under conditions of rest and exercise. *Percept. Psychophys.* **2,** 167–70.

Stone, H., Pryor, G. T. and Steinmetz, G. (1972). A comparison of olfactory adaptation among seven odorants and their relationship with several physico-chemical properties. *Percept. Psychophys.* **12,** 501–4.

Stuiver, M. (1958). *Biophysics of the sense of smell.* Thesis, University of Groningen.

Takhirov, M. T. (1969). Hygienic standards for acetic and acetic anhydride in air. *Hyg. Sanit. USSR*, **34** (4–6), 122–5.

Takhirov, M. T. (1974). Experimental study of the combined action of six atmospheric pollutants on the human organism (in Russian). *Gig. Sanit.* **39** (5), 100–2.

Tempelaar, H. C. G. (1913). *Over den invloed van licht op reukstoffen.* Thesis. University of Utrecht.

Teranishi, R., Buttery, R. G. and Guadagni, D. G. (1974). Odor quality and chemical structure in fruit and vegetable flavors. *Ann. NY Acad. Sci.* **23,** 209–16.

Thomas, M. D., Ivie, J. O., Abersold, J. N. and Hendricks, R. H. (1943). Automatic apparatus for determination of small concentrations of sulfur dioxide in air. *Ind. Eng. Chem. Anal. Ed.* **15,** 287–90.

Tkachev, P. G. (1963). Data for substantiating the maximum permissible concentration of aniline in the atmospheric air (in Russian). *Gig. Sanit.* **28**(4), 3–11.

Tkachev, P. G. (1969). Monoethylamine in the atmosphere: hygienic significance and standards. *Hyg. Sanit. USSR*, **34**(7–9), 149–53.

Tkachev, P. G. (1970). Triethylamine in air: hygienic significance and hygienic standards. *Hyg. Sanit. USSR*, **35**(10–12), 8–13.

Tkachev, P. G. (1978). Methods for determining threshold of olfactory sensation for chemical substances having an odour (in Russian). *Gig. Sanit.* **3,** 112–13.

Valentin, G. (1848). *Lehrbuch der Physiologie des Menschen* **2,** (2), 279–83. Friedrich Bieweg und Sohn, Braunschweig.

Valentin, G. (1850). *Lehrbuch der Physiologie des Menschen* **2,** (3), 271–4. Friedrich Bieweg und Sohn, Braunschweig.

Van Anrooij, A. (1931). Desodorisatie longs photochemischen weg. Thesis. University of Utrecht.

Van Gemert, L. J. (1984). *Compilation of odour threshold values in air.* Supplement V. Report No 84.220 TNO–CIVO Food Analysis Institute.

Van Gemert, L. J. and Nettenbreijer, A. H. (1977). *Compilation of odour threshold values in air and water.* Central Institute for Nutrition and Food Research TNO, Zeist, Netherlands, and National Institute for Water Supply, Voorburg, Netherlands.

Wilby, F. V. (1969). Variation in recognition odor threshold of a panel. *J. Air Pollut. Contr. Assoc.* **19,** 96–100.

Wolf, M. A., Rowe, V. K., McCollister, D. D., Hollingsworth, R. I. and Oyen, F. (1956). Toxicological studies of certain alkylated benzenes and benzene. *AMA Arch. Ind. Health*, **14,** 387–98.

Zwaardemaker, H. (1914). Geruch und Geschmack. In *Handbuch der physiologischen Methodik* III, **1,** (ed. R. Tigerstedt), pp.46–108. S. Hirzel, Leipzig (corrected by Tempelaar (1913)).

Appendix A: Mathematical procedure

1. Original data

The original data matrix represents 641 chemical compounds ×372 references, i.e. 238452 potential data. In fact, only 2075 threshold values are provided in these references; that is to say for each value present in the matrix, there are about 115 values absent.

The aim of this work consists of dividing the available data into two sets:

(a) standardized data gathered in the homogeneous set 'homoset';

(b) the remaining data, which constitutes the heterogeneous set 'heteroset'.

The process described below involves building the homoset from data within the initial global heterogeneous set.

In order to homogenize provided data, a stepwise process is defined to determinate gradually the weights for each reference and to extend the homoset. The general method (Patte *et al.* 1975) can be considered as a neobayesian stepwise method (Rouault and Capy 1986).

2. Initialization (Step 1)

Each reference r considers $n_r < 641$ substances. For each studied substance s, it offers a threshold value x_{rs}.

We choose the first arbitrary reference (Leonardos *et al.* 1969) according to the number of studied substances, the overlap with other references and a good consistency with other 'good' references (Patte *et al.* 1975).

For each substance s in the Leonardos reference, we have

$$X_s^{(1)} = x_{rs} \tag{1}$$

where X_s is the standardized threshold value (in negative log units, i.e. in p.ol (vol)) according to the set of initial data.

The first homoset$^{(1)}$ is made up with the Leonardos reference, and the first heteroset$^{(1)}$ with all the other references.

3. Reference weighting coefficient (Step i)

For each reference r from the homoset$^{(i)}$ or the heteroset$^{(i)}$, we consider the cardinal n_r of the intersection between the chemical compounds involved in this reference and those retained in the homoset$^{(i)}$.

If $n_r < 2$, the author is discarded at this step, i.e. it is put into the heteroset$^{(i+1)}$.

Otherwise, we calculate:

$$\Delta(\text{homoset}^{(i)},r) = [\sum_s (X_s^{(i)} - x_{rs})]/n_r. \qquad (2)$$

The weighting coefficient Δ estimates the systematic deviation between the reference r and the homoset$^{(i)}$.

4. Reference-homoset comparison (Step i)

With a reference r and the homoset$^{(i)}$, we can define an index of Profile Similarity PS:

$$\text{PS}(\text{homoset}^{(i)},r) = 1 - \sqrt{\sum_s [(X_s^{(i)} - x_{rs}) - \Delta(\text{homoset}^{(i)},r)]^2} / 7\sqrt{n_r} \qquad (3)$$

The highest value of PS (obtained by comparing identical profiles), is 1, according to the definition of formula 2. The lowest value depends on the maximal difference between $X_s^{(i)}$ and X_{rs} which can be obtained by chance. We considered this value to be 14 log units, which is the range of original data of the present compilation. Under these conditions, taking into account the constant 7 of equation 3, the lowest value is equal to -1.

A provisional table of significance was established by digital simulation, in the range of 14 log units (Table A.1).

Table A.1 Values of significance ($p \leqslant 0.01$) of PS index, according to formula 4, applied to olfactory thresholds

Number of substances	Profile similarity
n	PS (n, 0.01)
2	0.99
3	0.95
4	0.91
5	0.86
6	0.83
7	0.80
8	0.78
10	0.75
12	0.72
15	0.69
20	0.65
30	0.61
50	0.57
100	0.54

5. Making-up the homoset$^{(i+1)}$ (Step $i \rightarrow i+1$)

If the reference r is in the homoset$^{(i)}$ and PS(homoset$^{(i)}$,r) < PS(n, 0.01), this reference is discarded and is put in the heteroset$^{(i+1)}$. Otherwise, it is maintained in the homoset$^{(i+1)}$.

If the reference r is not in homoset$^{(i)}$ and PS(r,homoset$^{(i)}$) $\geqslant$ PS(n, 0.01), this reference is included in the new homoset$^{(i+1)}$. Otherwise, it is maintained in heteroset$^{(i+1)}$.

6. Stop test (Step $i \rightarrow i+1$)

If homoset$^{(i)}$ = homoset$^{(i+1)}$, then the process has converged and the statistics

$$X_s = X_s^{(i)} = X_s^{(i+1)} \tag{4}$$

$$\Delta r = \Delta(\text{homoset}^{(i)}, r) = \Delta(\text{homoset}^{(i+1)}, r) \tag{5}$$

can be considered as estimating the real values. If the initial reference (point 2) had not been chosen too 'marginal', the final homoset may be considered as independent from this arbitrary choice.

Otherwise, one more step must be taken. We define:

$$X_s^{(i+1)} = \sum_r [x_{rs} - \Delta(\text{homoset}^{(i)}, r)]/m \qquad (6)$$

where X_s is the standardized threshold value for a substance s and the m references in homoset$^{(i+1)}$. Then, the process restarts again at point 3.

Appendix B: Original and standardized threshold values

Original and weighted olfactory thresholds in air, according to Van Gemert *et al.* (1977, 1984) and presented in the same order, for 529 substances. The standard values proposed are the average of the weighted values. The units used: p.ol(mass) and p.ol(volume), are defined in the text (p.ol as abbreviation of 'olfactory power'). The global molecular formula, molecular weight and *Handbook of chemistry and physics* (50th edn) number are also reported (except for substances not given in this edition of the Handbook; in this case, a number was assigned using the Handbook system of nomenclature and indicated with an asterisk).

acetaldehyde see: **Ethanal**
acetamide see: **Ethanamide**

Acetic Acid ($C_2H_4O_2$ Mol.Wt.: 60.1 Handbook: a58)

	p.ol mass author	p.ol mass weighted	p.ol vol weighted
Passy 1893b, 1893c	5.15	5.18	5.58
Backman 1917	5.31	5.31	5.71
Grijns 1919	5.70	6.94	7.34
Hesse 1926	6.22	7.79	8.19
Mitsumoto 1926	6.69	7.73	8.13
Morimura 1934	5.73	6.90	7.30
Jung 1936	7.60	7.13	7.53
Stone 1963*c*	5.41	6.19	6.59
Stone and Bosley 1965	5.38	6.33	6.73
Takhirov 1974	6.22	5.41	5.81
Leonardos *et al.* 1969	5.60	5.59	5.99
Naus 1982	6.30	5.82	6.22
Punter 1983	7.04	7.44	7.84
Standard		6.44	6.84

Acetic anhydride ($C_4H_6O_3$ Mol.Wt.: 102.1 Handbook: a172)

	p.ol mass author	p.ol mass weighted	p.ol vol weighted
Takhirov 1969	3.55	5.50	6.13
Hellman and Small 1973, 1974	3.46	5.70	6.33
Standard		5.60	6.23

acetone see: **2-Propanone**
acetonitrile see: **Ethanenitrile**

Acetophenone (C_8H_8O Mol.Wt.: 120.2 Handbook: a645)

	p.ol mass author	p.ol mass weighted	p.ol vol weighted
Hellman and Small 1973, 1974	3.06	5.30	6.00
Gavaudan and Poussel 1966	3.88	6.18	6.88
Standard		5.74	6.44

acetyl chloride see: **Ethanoyl chloride**
acetylene see: **Ethyne**

1-Acetyl-4-methylbenzene (methyl acetophenone) ($C_9H_{10}O$ Mol.Wt.: 134.2 Handbook: a705)

	p.ol mass author	p.ol mass weighted	p.ol vol weighted
Appell 1969	7.06	7.43	8.18
Standard		7.43	8.18

acrolein see: **2-Propenal**
acrylic acid see: **Propenoic acid**
acrylonitrile see: **2-Propenenitrile**
allyl alcohol see: **2-Propen-1-ol**
allylamine see: **2-Propenylamine**
allylcarbylamine see: **2-Propenyl isocyanide**
allyl chloride see: **3-Chloropropene**

1-Allyl-2,5-dimethoxy-3,4-methylenedioxybenzene (apiole) ($C_{12}H_{14}O_4$ Mol.Wt.: 222.2 Handbook: a1368)

	p.ol mass author	p.ol mass weighted	p.ol vol weighted
Tempelaar 1913	0.45	3.80	4.77
Standard		3.80	4.77

allyl isothiocyanate see: **2-Propenyl isothiocyanate**

allyl mercaptan see: **2-Propene-1-thiol**

4-Allyl-2-methoxyphenol (eugenol) ($C_{10}H_{12}O_2$ Mol.Wt.: 164.2 Handbook: e611)

	p.ol mass author	p.ol mass weighted	p.ol vol weighted
Tempelaar 1913	3.87	7.22	8.06
Ohma 1922	3.87	6.81	7.65
Van Anrooij 1931	3.93	5.95	6.79
Baldus 1936	5.52	7.18	8.02
Jones 1955*b*	2.66	6.93	7.77
Appell 1969	8.33	8.70	9.54
Standard		7.14	7.97

1-Allyl-3,4-methylenedioxybenzene (safrole) ($C_{10}H_{10}O_2$ Mol.Wt.: 162.2 Handbook: s7)

	p.ol mass author	p.ol mass weighted	p.ol vol weighted
Tempelaar 1913	2.94	6.29	7.12
Ohma 1922	4.14	7.08	7.91
Van Anrooij 1931	6.70	8.72	9.55
Baldus 1936	5.54	7.20	8.03
Jones 1955*a*	4.34	7.25	8.08
Jones 1955*b*	2.76	7.03	7.86
Appell 1969	7.39	7.76	8.59
Köster 1971	1.90	6.16	6.99
Standard		7.19	8.02

1-Aminonaphthalene (1-naphthylamine) ($C_{10}H_9N$ Mol.Wt.: 143.2 Handbook: n36)

	p.ol mass author	p.ol mass weighted	p.ol vol weighted
Backman 1917	3.93	6.69	7.47
Standard		6.69	7.47

2-Aminonaphthalene (2-naphthylamine) ($C_{10}H_9N$ Mol.Wt.: 143.2 Handbook: n38)

	p.ol mass author	p.ol mass weighted	p.ol vol weighted
Backman 1917	3.02	5.78	6.56
Standard		5.78	6.56

Ammonia (NH_3 Mol.Wt.: 17.0 Handbook: I, a76)

	p.ol mass author	p.ol mass weighted	p.ol vol weighted
Valentin 1848, 1850	4.68	5.82	5.67
Fieldner *et al.* 1921	4.43	5.38	5.23
Smolczyk and Cobler 1930	5.65	5.86	5.71
Geier 1936	5.90	5.48	5.33
Patty 1962*a*	5.46	5.42	5.27
Leonardos *et al.* 1969	4.48	4.47	4.32
Stephens 1971	5.57	5.48	5.33
Nishida *et al.* 1975	5.08	5.23	5.08
Logtenberg 1978	5.29	4.76	4.61
Anon 1980	7.00	6.03	5.88
Naus 1982	5.83	5.35	5.20
Standard		5.39	5.24

amyl acetate see: **Pentyl acetate**

amyl alcohol see: **1-Pentanol**

tert.amyl alcohol see: **2-Methylbutan-2-ol**

amyl isovalerate see: **Pentyl 3-methylbutanoate**

amyl mercaptan see: **1-Pentanethiol**

tert.amyl mercaptan see: **2-Methylbutane-2-thiol**

amyl salicylate see: **Pentyl salicylate**

5-α-Androst-16-en-3-one ($C_{19}H_{28}O$ Mol.Wt.: 272.4 Handbook: a992*)

	p.ol mass author	p.ol mass weighted	p.ol vol weighted
Amoore 1977	8.67	7.91	8.97
Standard		7.91	8.97

anethole see: **1-Methoxy-4-propenylbenzene**

Aniline (C_6H_7N Mol.Wt.: 93.1 Handbook: a1002)

	p.ol mass author	p.ol mass weighted	p.ol vol weighted
Tempelaar 1913	6.01	6.60	7.19
Backman 1917	5.27	5.27	5.86
Huijer 1918	4.34	5.40	5.99
Geier 1936	5.87	5.45	6.04
Tkachev 1963	6.43	5.68	6.27
Leonardos *et al.* 1969	5.42	5.41	6.00
Naus 1982	5.70	5.22	5.81
Standard		5.58	6.17

anisaldehyde see: **4-Methoxybenzaldehyde**

anisole see: **Methoxybenzene**

apiole see: **1-Allyl-2,5-dimethoxy-3,4-methylenedioxybenzene**

Arsine (AsH_3 Mol.Wt.: 78.0 Handbook: I, a298)

	p.ol mass author	p.ol mass weighted	p.ol vol weighted
Patty 1926*b*	5.50	5.46	5.97
Standard		5.46	5.97

artificial musk see: **1-tert.Butyl-3-methyl-2,4,6-trinitrobenzene**

3-Aza-2,4-dimethylpentane-2,4-diol (diisopropanolamine) ($C_6H_{15}NO_2$ Mol.Wt.: 133.2 Handbook: a895)

	p.ol mass author	p.ol mass weighted	p.ol vol weighted
England *et al.* 1978	7.16	8.21	8.95
Standard		8.21	8.95

3-Azapentane-1,5-diol (diethanolamine) ($C_4H_{11}NO_2$ Mol.Wt.: 105.1 Handbook: a830)

	p.ol mass author	p.ol mass weighted	p.ol vol weighted
England *et al.* 1978	5.92	6.97	7.61
Standard		6.97	7.61

azine see: **Pyridine**
azolidine see: **Pyrrolidine**

Benzaldehyde (C_7H_6O Mol.Wt.: 106.1 Handbook: b35)

	p.ol mass author	p.ol mass weighted	p.ol vol weighted
Backman 1917	6.39	6.39	7.04
Rocen 1920	5.77	6.32	6.97
Ohma 1922	6.35	6.53	7.18
Katz and Talbert 1930	6.74	6.94	7.59
Jones 1955*b*	5.38	6.89	7.54
Pliska and Janicek 1965	4.88	7.68	8.33
Knuth 1973	6.56	6.33	6.98
Laing 1975	5.36	6.74	7.39
Standard		6.73	7.38

Benzene (C_6H_6 Mol.Wt.: 78.1 Handbook: b202)

	p.ol mass author	p.ol mass weighted	p.ol vol weighted
Backman 1917	5.17	5.17	5.68
Backman 1918	5.28	4.80	5.31
Grijns 1919	3.38	4.62	5.13
Schley 1934	5.06	4.69	5.20
Jones 1954	3.31	4.99	5.50
Jones 1955*b*	3.75	5.26	5.77
Deadman and Prigg 1959	5.05	4.73	5.24
Naus 1962	5.22	4.60	5.11
Gusev 1965	5.48	5.09	5.60
May 1966	3.75	4.94	5.45
Elfimova 1966	5.60	4.89	5.40
Leonardos *et al.* 1969	4.83	4.82	5.33
Dravnieks and O'Donnell 1971	4.42	4.89	5.40
Köster 1971	4.43	5.93	6.44
Dravnieks and Laffort 1972	4.49	4.59	5.10
Laffort and Dravnieks 1973	4.84	5.25	5.76
Dravnieks 1974	3.42	4.62	5.13
Naus 1982	5.83	5.35	5.86
Punter 1983	3.97	4.37	4.88
Standard		4.92	5.44

Benzenethiol (thiophenol) (C_6H_6S Mol.Wt.: 110.2 Handbook: b854)

	p.ol mass author	p.ol mass weighted	p.ol vol weighted
Katz and Talbert 1930	8.92	9.12	9.78
Stuiver 1958	9.86	8.57	9.23
Standard		8.84	9.51

1,4-Benzoquinone (quinone) ($C_6H_4O_2$ Mol.Wt.: 108.1 Handbook: b2069)

	p.ol mass author	p.ol mass weighted	p.ol vol weighted
Backman 1917	7.28	7.28	7.97
Standard		7.28	7.97

Benzoyl chloride (C_7H_5OCl Mol.Wt.: 140.6 Handbook: b1268)

	p.ol mass author	p.ol mass weighted	p.ol vol weighted
Schley 1934	7.77	7.40	8.17
Standard		7.40	8.17

Benzyl acetate ($C_9H_{10}O_2$ Mol.Wt.: 150.2 Handbook: a173)

	p.ol mass author	p.ol mass weighted	p.ol vol weighted
Appell 1969	9.00	6.61	7.41
Köster 1971	3.97	5.47	6.27
Standard		6.04	6.84

benzylbenzene see: **Diphenylmethane**

Benzyl chloride (α-chlorotoluene) (C_7H_7Cl Mol.Wt.: 126.6 Handbook: t391)

	p.ol mass author	p.ol mass weighted	p.ol vol weighted
Katz and Talbert 1930	6.68	6.88	7.60
Leonardos *et al.* 1969	6.62	6.61	7.33
Standard		6.74	7.47

benzyl mercaptan see: α-**Toluenethiol**

Benzylthiotoluene (dibenzyl sulfide) ($C_{14}H_{14}S$ Mol.Wt.: 214.3 Handbook: s294)

	p.ol mass author	p.ol mass weighted	p.ol vol weighted
Katz and Talbert 1930	7.28	7.48	8.43
Leonardos *et al.* 1969	7.75	7.74	8.69
Standard		7.61	8.56

Borneol ($C_{10}H_{18}O$ Mol.Wt.: 154.3 Handbook: b2414)

	p.ol mass author	p.ol mass weighted	p.ol vol weighted
Tempelaar 1913	7.28	7.87	8.68
Standard		7.87	8.68

Bornyl acetate ($C_{12}H_{20}O_2$ Mol.Wt.: 196.3 Handbook: b2417)

	p.ol mass author	p.ol mass weighted	p.ol vol weighted
Tempelaar 1913	6.36	6.95	7.86
Standard		6.95	7.86

Bornyl salicylate ($C_{17}H_{22}O_3$ Mol.Wt.: 274.4 Handbook: b2422*)

	p.ol mass author	p.ol mass weighted	p.ol vol weighted
Baldus 1936	7.52	6.42	7.48
Standard		6.42	7.48

Bromine (Br_2 Mol.Wt.: 159.8 Handbook: I, b214)

	p.ol mass author	p.ol mass weighted	p.ol vol weighted
Valentin 1848, 1850	5.53	6.67	7.49
Henning 1924	6.70	5.89	6.71
Leonardos *et al.* 1969	6.53	6.52	7.34
Standard		6.36	7.18

α-bromoacetophenone see: **Phenacyl bromide**

Bromobenzene (C_6H_5Br Mol.Wt.: 157.0 Handbook: b309)

	p.ol mass author	p.ol mass weighted	p.ol vol weighted
Backman 1917	5.72	5.72	6.54
Punter 1983	5.92	6.32	7.14
Standard		6.02	6.84

Bromoethane (C_2H_5Br Mol.Wt.: 109.0 Handbook: e195)

	p.ol mass author	p.ol mass weighted	p.ol vol weighted
Backman 1917	4.85	4.85	5.51
Standard		4.85	5.51

bromoform see: **Tribromomethane**

1-Bromo-2-methylbenzene (C_7H_7Br Mol.Wt.: 171.0 Handbook: t357)

	p.ol mass author	p.ol mass weighted	p.ol vol weighted
Backman 1917	6.96	6.96	7.81
Standard		6.96	7.81

1-(2-Bromo-3-methylbutanoyl)urea (bromural) ($C_6H_{11}O_2Br$ Mol.Wt.: 223.1 Handbook: u38)

	p.ol mass author	p.ol mass weighted	p.ol vol weighted
Backman 1917	5.54	5.54	6.51
Standard		5.54	6.51

2-Bromophenol (C_6H_5BrO Mol.Wt.: 173.0 Handbook: p592)

	p.ol mass author	p.ol mass weighted	p.ol vol weighted
Kendall *et al.* 1968	11.15	10.74	11.60
Standard		10.74	11.60

bromural see: **1-(2-Bromo-3-methylbutanoyl)urea**

1,3-Butadiene (C_4H_6 Mol.Wt.: 54.1 Handbook: b2436)

	p.ol mass author	p.ol mass weighted	p.ol vol weighted
Mullins 1955	3.78	5.53	5.88
Deadman and Prigg 1959	5.68	5.36	5.71
Laffort and Dravnieks 1973	5.24	5.65	6.00
Hellman and Small 1974	6.00	5.48	5.83
Standard		5.50	5.86

Butanal (C_4H_8O Mol.Wt.: 72.1 Handbook: b2471)

	p.ol mass author	p.ol mass weighted	p.ol vol weighted
Backman 1917	7.88	7.87	8.36
Pliska and Janicek 1965	4.82	7.61	8.10
Hellman and Small 1973, 1974	7.88	7.35	7.84
Teranishi *et al.* 1974	7.37	7.27	7.76
Anon 1980	9.07	8.09	8.58
Hall and Andersson 1983	6.70	7.16	7.65
Standard		7.56	8.05

Butane (C_4H_{10} Mol.Wt.: 58.1 Handbook: b2485)

	p.ol mass author	p.ol mass weighted	p.ol vol weighted
Patty and Yant 1929	1.93	3.03	3.41
Mullins 1955	2.21	3.96	4.34
Laffort and Dravnieks 1973	2.53	2.94	3.32
Standard		3.31	3.69

2,3-Butanedione (diacetyl) ($C_4H_6O_2$ Mol.Wt.: 86.1 Handbook: b2632)

	p.ol mass author	p.ol mass weighted	p.ol vol weighted
Backman 1917	8.38	8.38	8.93
Van Anrooij 1931	8.61	7.87	8.42
Appell 1969	8.59	6.20	6.75
Hall and Andersson 1983	8.31	8.78	9.33
Standard		7.80	8.36

1-Butanethiol ($C_4H_{10}S$ Mol.Wt.: 90.2 Handbook: b2643)

	p.ol mass author	p.ol mass weighted	p.ol vol weighted
Allison and Katz 1919	4.74	7.32	7.90
Katz and Talbert 1930	8.43	8.63	9.21
Deadman and Prigg 1959	8.82	8.50	9.08
Blinova 1965	7.77	8.34	8.92
Kniebes *et al.* 1969	8.52	8.32	8.90
Wilby 1969	8.56	8.48	9.06
Standard		8.27	8.85

2-Butanethiol (sec.butyl mercaptan) ($C_4H_{10}S$ Mol.Wt.: 90.2 Handbook: b2645)

	p.ol mass author	p.ol mass weighted	p.ol vol weighted
Stuiver 1958	10.69	9.40	9.98
Blanchard 1976	8.15	8.92	9.50
Standard		9.16	9.74

Butanoic acid (butyric acid) ($C_4H_8O_2$ Mol.Wt.: 88.1 Handbook: b2651)

	p.ol mass author	p.ol mass weighted	p.ol vol weighted
Passy 1893*b*, 1893*c*	9.00	9.03	9.60
Backman 1917	7.69	7.69	8.26
Allison and Katz 1919	5.04	7.62	8.19
Hesse 1926	7.00	8.57	9.14
Mitsumoto 1926	6.78	7.82	8.39
Hesse 1927	5.82	7.44	8.01
Jones 1955*a*	7.79	7.94	8.51
Jones 1955*b*	5.52	7.03	7.60
Kerka and Humphreys 1956	8.69	8.31	8.88
Nader 1958	7.23	6.98	7.55
Bocca and Battiston 1964	7.50	7.80	8.37
Guadagni 1966	6.61	6.37	6.94
Goldenberg 1967	9.39	8.16	8.73
Leonardos *et al.* 1969	8.39	8.38	8.95
Cormack *et al.* 1974	8.39	8.02	8.59
Bedborough and Trott 1979	8.06	7.36	7.93
Anon 1980	9.45	8.48	9.05
Punter 1983	7.69	8.09	8.66
Standard		7.84	8.41

1-Butanol ($C_4H_{10}O$ Mol.Wt.: 74.1 Handbook: b2864)

	p.ol mass author	p.ol mass weighted	p.ol vol weighted
Passy 1892*c*	6.00	6.15	6.64
Backman 1917	6.34	6.34	6.83
Jung 1936	6.65	6.18	6.67
Mullins 1955	4.43	6.18	6.67
Jones 1955*a*	5.51	5.66	6.15
Jones 1955*b*	4.38	5.89	6.38
Scherberger *et al.* 1958	4.35	5.77	6.26
Janicek *et al.* 1960	4.70	6.28	6.77
Naus 1962	5.40	4.78	5.27
Pliska and Janicek 1965	2.52	5.32	5.81
Gavaudan and Poussel 1966	5.96	5.50	5.99
May 1966	4.48	5.67	6.16
Cain 1969	4.22	5.77	6.26
Corbit and Engen 1971	4.79	5.75	6.24
Dravnieks and Laffort 1972	5.00	5.10	5.59
Laffort and Dravnieks 1973	6.05	6.46	6.95
Hellman and Small 1974	6.05	5.53	6.02

1-Butanol *continued*

Laing *et al.* 1978	4.98	6.10	6.59
Laing 1982	5.52	6.17	6.66
Punter 1983	5.43	5.83	6.32
Standard		5.82	6.31

2-Butanol (sec.butyl alcohol) ($C_4H_{10}O$ Mol.Wt.: 74.1
Handbook: b2866)

	p.ol mass author	p.ol mass weighted	p.ol vol weighted
Jung 1936	5.13	4.66	5.15
Jones 1955*b*	4.10	5.61	6.10
Laffort and Dravnieks 1973	5.05	5.46	5.95
Hellman and Small 1974	6.40	5.88	6.37
Bedborough and Trott 1979	5.48	4.78	5.27
Standard		5.28	5.77

(d)2-Butanol (sec.butyl alcohol) ($C_4H_{10}O$ Mol.Wt.: 74.1
Handbook: b2865)

	p.ol mass author	p.ol mass weighted	p.ol vol weighted
Punter 1983	4.23	4.63	5.12
Standard		4.63	5.12

(l)2-Butanol (sec.butyl alcohol) ($C_4H_{10}O$ Mol.Wt.: 74.1
Handbook: b2867)

	p.ol mass author	p.ol mass weighted	p.ol vol weighted
Punter 1983	4.38	4.78	5.27
Standard		4.78	5.27

2-Butanone (methyl ethyl ketone) (C_4H_8O Mol.Wt.: 72.1 Handbook: b2912)

	p.ol mass author	p.ol mass weighted	p.ol vol weighted
Backman 1917	4.18	4.18	4.66
May 1966	4.09	5.28	5.76
Leonardos *et al.* 1969	4.54	4.53	5.01
Dravnieks and Laffort 1972	4.65	4.75	5.23
Dravnieks 1974	3.60	4.80	5.28
Hellman and Small 1974	5.23	4.71	5.19
Anon 1980	5.07	4.10	4.58
Hall and Andersson 1983	4.21	4.68	5.16
Standard		4.63	5.11

trans-2-Butenal (crotonaldehyde) (C_4H_6O Mol.Wt.: 70.1 Handbook: b2933)

	p.ol mass author	p.ol mass weighted	p.ol vol weighted
Katz and Talbert 1930	6.49	6.69	7.16
Teranishi *et al.* 1974	6.37	6.28	6.75
Hall and Andersson 1983	5.77	6.24	6.71
Standard		6.41	6.87

1-Butene (C_4H_8 Mol.Wt.: 56.1 Handbook: b2940)

	p.ol mass author	p.ol mass weighted	p.ol vol weighted
Katz and Talbert 1930	5.68	5.88	6.25
Mullins 1955	4.41	6.16	6.53
Knuth 1973	5.92	5.69	6.06
Standard		5.91	5.71

2-Butene (C_4H_8 Mol.Wt.: 56.1 Handbook: b2942)

	p.ol mass author	p.ol mass weighted	p.ol vol weighted
Katz and Talbert 1930	5.32	5.52	5.89
Mullins 1955	2.57	4.32	4.69
Standard		4.92	5.29

cis-2-Butene (C_4H_8 Mol.Wt.: 56.1 Handbook: b2941)

	p.ol mass author	p.ol mass weighted	p.ol vol weighted
Mullins 1955	4.54	6.29	6.66
Standard		6.29	6.66

2-Butene-1-thiol (crotyl mercaptan) (C_4H_8S Mol.Wt.: 88.2 Handbook: b3006*)

	p.ol mass author	p.ol mass weighted	p.ol vol weighted
Katz and Talbert 1930	9.11	9.31	9.88
Standard		9.31	9.88

Butoxybutane (butyl ether) ($C_8H_{18}O$ Mol.Wt.: 130.2 Handbook: e471)

	p.ol mass author	p.ol mass weighted	p.ol vol weighted
Dravnieks and Laffort 1972	7.53	7.63	8.36
Dravnieks 1974	5.10	6.30	7.03
Hellman and Small 1974	6.40	5.88	6.61
Standard		6.60	7.33

2-Butoxyethanol (butyl cellosolve) ($C_6H_{14}O_2$ Mol.Wt.: 118.2 Handbook: e348)

	p.ol mass author	p.ol mass weighted	p.ol vol weighted
Hellman and Small 1973, 1974	6.30	5.78	6.47
Standard		5.78	6.47

2-Butoxyethyl acetate (butyl cellosolve acetate) ($C_8H_{16}O_3$ Mol.Wt.: 160.2 Handbook: a177*)

	p.ol mass author	p.ol mass weighted	p.ol vol weighted
Hellman and Small 1973, 1974	6.16	5.64	6.46
Standard		5.64	6.46

Butyl acetate ($C_6H_{12}O_2$ Mol.Wt.: 116.2 Handbook: a178)

	p.ol mass author	p.ol mass weighted	p.ol vol weighted
Backman 1917	5.82	5.82	6.51
Jung 1936	7.35	6.88	7.57
Scherberger *et al.* 1958	4.01	5.43	6.12
Gofmekler 1960	6.22	5.67	6.36
Naus 1962	6.15	5.53	6.22
May 1966	4.45	5.64	6.33
Köster 1971	3.03	4.53	5.22
Dravnieks and Laffort 1972	7.39	7.49	8.18
Dravnieks 1974	5.52	6.72	7.41
Hellman and Small 1974	7.52	7.00	7.69
Anon 1980	6.49	5.52	6.21
Standard		6.03	6.71

butyl acrylate see: **Butyl 2-propenoate**

sec.butyl alcohol see: **2-Butanol**

tert.butyl alcohol see: **2-Methylpropan-2-ol**

Butylamine ($C_4H_{11}N$ Mol.Wt.: 73.1 Handbook: b2486)

	p.ol mass author	p.ol mass weighted	p.ol vol weighted
Hellman and Small 1973, 1974	6.62	6.10	6.58
Laing *et al.* 1978	4.38	5.50	5.98
Standard		5.80	6.28

butyl cellosolve see: **2-Butoxyethanol**

butyl cellosolve acetate see: **2-Butoxyethyl acetate**

1-tert.Butyl-3,5-dimethyl-2,4,6-trinitrobenzene (musk xylene) ($C_{12}H_{15}N_3O_6$ Mol.Wt.: 297.3 Handbook: b383)

	p.ol mass author	p.ol mass weighted	p.ol vol weighted
Backman 1917	8.05	8.05	9.14
Allison and Katz 1919	7.40	9.98	11.07
Katz and Talbert 1930	9.23	9.43	10.52
Standard		9.15	10.24

1-tert.Butyl-3,5-dinitro-2-methoxy-4-methylbenzene (musk ambrette) ($C_{12}H_{16}N_2O_5$ Mol.Wt.: 268.3 Handbook: b385)

	p.ol mass author	p.ol mass weighted	p.ol vol weighted
Appell 1969	13.04	10.65	11.70
Standard		10.65	11.70

butyl ether see: **Butoxybutane**

Butyl formate ($C_5H_{10}O_2$ Mol.Wt.: 102.1 Handbook: f132)

	p.ol mass author	p.ol mass weighted	p.ol vol weighted
May 1966	4.15	5.34	5.97
Standard		5.34	5.97

sec.butyl mercaptan see: **2-Butanethiol**

tert.butyl mercaptan see: **2-Methylpropane-2-thiol**

1-tert.Butyl-3-methyl-2,4,6-trinitrobenzene (artificial musk) ($C_{11}H_{13}N_3O_6$ Mol.Wt.: 283.2 Handbook: b404)

	p.ol mass author	p.ol mass weighted	p.ol vol weighted
Passy 1892*a*, 1892*b*	12.65	12.80	13.87
Zwaardemaker 1914	9.00	8.92	9.99
Standard		10.86	11.93

Butyl 2-propenoate (butyl acrylate) ($C_7H_{12}O_2$ Mol.Wt.: 128.2 Handbook: p1770)

	p.ol mass author	p.ol mass weighted	p.ol vol weighted
Anon 1980	8.82	7.85	8.58
Standard		7.85	8.58

Butyl salicylate ($C_{11}H_{14}O_3$ Mol.Wt.: 194.2 Handbook: b1658)

	p.ol mass author	p.ol mass weighted	p.ol vol weighted
Baldus 1936	8.23	7.13	8.04
Standard		7.13	8.04

Butylthiobutane (dibutyl sulfide) ($C_8H_{18}S$ Mol.Wt.: 146.3 Handbook: s295)

	p.ol mass author	p.ol mass weighted	p.ol vol weighted
Katz and Talbert 1930	7.04	7.24	8.03
Standard		7.24	8.03

2-Butyne (C_6H_6 Mol.Wt.: 54.1 Handbook: b3071)

	p.ol mass author	p.ol mass weighted	p.ol vol weighted
Mullins 1955	3.16	4.91	5.26
Standard		4.91	5.26

butyric acid see: **Butanoic acid**

Camphor ($C_{10}H_{16}O$ Mol.Wt.: 152.2 Handbook: c32)

	p.ol mass author	p.ol mass weighted	p.ol vol weighted
Passy 1892*a*, 1892*b*	5.30	5.45	6.25
Zwaardemaker 1914	7.80	7.72	8.52
Backman 1917	6.09	6.09	6.89
Ohma 1922	7.22	7.40	8.20
Mitsumoto 1926	4.85	5.89	6.69
Morimura 1934	5.21	6.38	7.18
Standard		6.49	7.29

capric acid see: **Decanoic acid**
caproic acid see: **Hexanoic acid**
caprylic acid see: **Octanoic acid**
carbitol see: **3,6-Dioxaoctan-1-ol**
carbitol acetate see: **3,6-Dioxaoctan-1-ylacetate**

Carbon disulfide (CS_2 Mol.Wt.: 76.1 Handbook: c161)

	p.ol mass author	p.ol mass weighted	p.ol vol weighted
Deadman and Prigg 1959	7.16	6.84	7.34
Baikov 1963	6.70	6.54	7.04
Leonardos *et al.* 1969	6.19	6.18	6.68
Naus 1982	7.00	6.52	7.02
Standard		6.52	7.02

carbon tetrachloride see: **Tetrachloromethane**

Carbonyl chloride (phosgene) ($COCl_2$ Mol.Wt.: 98.9 Handbook: p732)

	p.ol mass author	p.ol mass weighted	p.ol vol weighted
Fieldner *et al.* 1921	4.63	5.58	6.20
Schley 1934	6.30	5.93	6.55
Patty 1962*c*	5.69	5.65	6.27
Leonardos *et al.* 1969	5.39	5.38	6.00
Standard		5.64	6.26

carvacrol see: **5-Isopropyl-2-methylphenol**

Carvone ($C_{10}H_{14}O$ Mol.Wt.: 150.2 Handbook: c212)

	p.ol mass author	p.ol mass weighted	p.ol vol weighted
Tempelaar 1913	6.26	6.85	7.65
Standard		6.85	7.65

cellosolve see: **2-Ethoxyethanol**

cellosolve acetate see: **2-Ethoxyethyl acetate**

chloral see: **Trichloroethanal**

chloral hydrate see: **Trichloroethanal hydrate**

Chlorine (C_{12} Mol.Wt.: 70.9 Handbook: I, c369)

	p.ol mass author	p.ol mass weighted	p.ol vol weighted
Fieldner *et al.* 1921	5.00	5.95	6.42
Smolczyk and Cobles 1930	5.35	5.56	6.03
Beck 1959	6.67	6.40	6.87
Leonardos *et al.* 1969	6.22	6.21	6.68
Naus 1982	5.52	5.04	5.51
Standard		5.83	6.30

2-Chloroacetic acid ($C_2H_3ClO_2$ Mol.Wt.: 94.5 Handbook: a309)

	p.ol mass author	p.ol mass weighted	p.ol vol weighted
Backman 1917	6.22	6.22	6.82
Smith and Hochstettler 1969	7.30	6.86	7.46
Standard		6.54	7.14

α-chloroacetophenone see: **Phenacyl chloride**

2-Chloroaniline (C_6H_6ClN Mol.Wt.: 127.6 Handbook: a1043)

	p.ol mass author	p.ol mass weighted	p.ol vol weighted
Punter 1983	6.16	6.56	7.29
Standard		6.56	7.29

3-Chloroaniline (C_6H_6ClN Mol.Wt.: 127.6 Handbook: a1046)

	p.ol mass author	p.ol mass weighted	p.ol vol weighted
Punter 1983	5.95	6.35	7.08
Standard		6.35	7.08

4-Chloroaniline (C_6H_6ClN Mol.Wt.: 127.6 Handbook: a1048)

	p.ol mass author	p.ol mass weighted	p.ol vol weighted
Punter 1983	5.80	6.20	6.93
Standard		6.20	6.93

Chlorobenzene (C_6H_5Cl Mol.Wt.: 112.6 Handbook: b407)

	p.ol mass author	p.ol mass weighted	p.ol vol weighted
Backman 1917	5.11	5.11	5.78
Leonardos *et al.* 1969	6.01	6.00	6.67
Smith and Hochstettler 1969	5.52	5.08	5.75
Punter 1983	5.23	5.63	6.30
Standard		5.45	6.13

1-Chlorobutane (C_4H_9Cl Mol.Wt.: 92.6 Handbook: b2511)

	p.ol mass author	p.ol mass weighted	p.ol vol weighted
Mullins 1955	3.34	5.09	5.68
Hellman and Small 1973, 1974	4.48	3.96	4.55
Standard		4.53	5.12

1-Chloro-2-(1-chloroisopropoxy)propane (dichloroisopropylether) ($C_6H_{12}Cl_2O$ Mol.Wt.: 171.1 Handbook: e498)

	p.ol mass author	p.ol mass weighted	p.ol vol weighted
Punter 1983	6.59	6.99	7.84
Standard		6.99	7.84

1-Chlorodecane ($C_{10}0H_{21}Cl$ Mol.Wt.: 176.7 Handbook: d25)

	p.ol mass author	p.ol mass weighted	p.ol vol weighted
Mullins 1955	4.66	6.41	7.28
Standard		6.41	7.28

Chloroethane (C_2H_5Cl Mol.Wt.: 64.5 Handbook: e199)

	p.ol mass author	p.ol mass weighted	p.ol vol weighted
Backman 1917	4.96	4.96	5.39
Standard		4.96	5.39

1-Chloro-1-(chloroethylthio)ethane (1,1'-dichlorodiethyl sulfide) ($C_4H_8Cl_2S$ Mol.Wt.: 159.1 Handbook: s307)

	p.ol mass author	p.ol mass weighted	p.ol vol weighted
Katz and Talbert 1930	7.83	8.03	8.85
Standard		8.03	8.85

chloroform see: **Trichloromethane**

1-Chloroheptane ($C_7H_{15}Cl$ Mol.Wt.: 134.7 Handbook: h98)

	p.ol mass author	p.ol mass weighted	p.ol vol weighted
Mullins 1955	3.80	5.55	6.30
Standard		5.55	6.30

Chloromethane (CH_3Cl Mol.Wt.: 50.5 Handbook: a228)

	p.ol mass author	p.ol mass weighted	p.ol vol weighted
Leonardos *et al.* 1969	4.68	4.67	4.99
Standard		4.67	4.99

5-Chloro-2-methylaniline (C_7H_8ClN Mol.Wt.: 141.6 Handbook: t302)

	p.ol mass author	p.ol mass weighted	p.ol vol weighted
Punter 1983	7.00	7.40	8.17
Standard		7.40	8.17

1-Chloro-2-methylbenzene (C_7H_7Cl Mol.Wt.: 126.6 Handbook: t391)

	p.ol mass author	p.ol mass weighted	p.ol vol weighted
Backman 1917	5.94	5.94	6.66
Standard		5.94	6.66

2-Chlorophenol (C_6H_5ClO Mol.Wt.: 128.6 Handbook: p613)

	p.ol mass author	p.ol mass weighted	p.ol vol weighted
Katz and Talbert 1930	7.72	7.92	8.65
Kendall *et al.* 1968	9.30	8.89	9.62
Standard		8.41	9.14

4-Chlorophenol (C_6H_5ClO Mol.Wt.: 128.6 Handbook: p615)

	p.ol mass author	p.ol mass weighted	p.ol vol weighted
Kendall *et al.* 1968	9.00	8.59	9.32
Standard		8.59	9.32

3-Chloropropene (allyl chloride) (C_3H_5Cl Mol.Wt.: 76.5 Handbook: p1722)

	p.ol mass author	p.ol mass weighted	p.ol vol weighted
Leonardos *et al.* 1969	5.83	5.82	6.32
Standard		5.82	6.32

α-chlorotoluene see: **Benzyl chloride**

1,8-Cineole (eucalyptol) ($C_{10}H_{18}O$ Mol.Wt.: 154.3 Handbook: c335)

	p.ol mass author	p.ol mass weighted	p.ol vol weighted
Tempelaar 1913	7.26	7.85	8.66
Grijns 1919	5.15	6.39	7.20
Ohma 1922	6.72	6.90	7.71
Baldus 1936	8.52	7.42	8.23
Appell 1969	7.85	5.46	6.27
Randebrock 1971	9.70	7.85	8.66
Standard		6.98	7.79

cinnamaldehyde see: **3-Phenylpropenal**
cinnamic alcohol see: **3-Phenylprop-2-en-1-ol**
citral see: **3,7-Dimethylocta-2,6-dienal**

Citral a (geranial) ($C_{10}H_{16}O$ Mol.Wt.: 152.2 Handbook: c425)

	p.ol mass author	p.ol mass weighted	p.ol vol weighted
Bocca and Battiston 1964	7.40	7.70	8.50
Standard		7.70	8.50

citronellal see: **3,7-Dimethyloct-6-enal**
citronellol see; **3,7-Dimethyloct-6-en-l-ol**

Coumarin ($C_9H_6O_2$ Mol.Wt.: 146.2 Handbook: c526)

	p.ol mass author	p.ol mass weighted	p.ol vol weighted
Passy 1892*a*, 1892*b*	7.66	7.81	8.59
Tempelaar 1913	9.11	9.70	10.48
Backman 1917	7.60	7.60	8.38
Katz and Talbert 1930	7.70	7.90	8.68
Appell 1969	11.16	8.77	9.55
Standard		8.35	9.14

m-cresol see: **3-Methylphenol**
o-cresol see: **2-Methylphenol**

p-cresol see: **4-Methylphenol**

p-cresyl acetate see: **4-Methylphenyl acetate**

crotonaldehyde see: **trans-2-Butenal**

crotyl mercaptan see: **2-Butene-1-thiol**

cumene see: **Isopropylbenzene**

Cyclohexane (C_6H_{12} Mol.Wt.: 84.2 Handbook: c637)

	p.ol mass author	p.ol mass weighted	p.ol vol weighted
Schley 1934	4.40	4.03	4.58
Jones 1955*b*	3.04	4.55	5.10
Dravnieks and Laffort 1972	3.50	3.60	4.15
Stone *et al.* 1972	4.44	4.52	5.07
Laffort and Dravnieks 1973	3.78	4.19	4.74
Dravnieks 1974	2.56	3.76	4.31
Standard		4.11	4.66

Cyclohexanol ($C_6H_{12}O$ Mol.Wt.: 100.2 Handbook: c742)

	p.ol mass author	p.ol mass weighted	p.ol vol weighted
Punter 1983	6.19	6.59	7.21
Standard		6.59	7.21

Cyclohexanone ($C_6H_{10}O$ Mol.Wt.: 98.2 Handbook: c785)

	p.ol mass author	p.ol mass weighted	p.ol vol weighted
Stone *et al.* 1967	5.94	5.67	6.28
Köster 1971	3.08	4.58	5.19
Stone *et al.* 1972	5.80	5.88	6.49
Hellman and Small 1973, 1974	6.32	5.80	6.41
Laing 1975	4.40	5.78	6.39
Standard		5.54	6.15

Cyclohexene (C_6H_{10} Mol.Wt.: 82.2 Handbook: c811)

	p.ol mass author	p.ol mass weighted	p.ol vol weighted
Deadman and Prigg 1959	6.23	5.91	6.44
Standard		5.91	6.44

Cyclooctane (C_8H_{16} Mol.Wt.: 112.2 Handbook: c855)

	p.ol mass author	p.ol mass weighted	p.ol vol weighted
Stone *et al.* 1972	5.44	5.52	6.19
Standard		5.52	6.19

Cyclopentadiene (C_5H_6 Mol.Wt.: 66.1 Handbook: c865)

	p.ol mass author	p.ol mass weighted	p.ol vol weighted
Deadman and Prigg 1959	5.30	4.98	5.42
Standard		4.98	5.42

Cyclopentanol ($C_5H_{10}O$ Mol.Wt.: 86.1 Handbook: c892)

	p.ol mass author	p.ol mass weighted	p.ol vol weighted
Köster 1971	2.21	3.71	4.27
Standard		3.71	4.27

Cyclopentanone (C_5H_8O Mol.Wt.: 84.1 Handbook: c894)

	p.ol mass author	p.ol mass weighted	p.ol vol weighted
Köster 1971	3.73	5.23	5.77
Standard		5.23	5.77

o-cymene see: **1-Isopropopyl-2-methylbenzene**

p-cymene see: **1-Isopropopyl-4-methylbenzene**

trans-2,trans-4-Decadienal ($C_{10}H_{16}O$ Mol.Wt.: 152.3 Handbook: d1*)

	p.ol mass author	p.ol mass weighted	p.ol vol weighted
Teranishi *et al.* 1974	9.75	9.66	10.46
Standard		9.66	10.46

Decanal ($C_{10}H_{20}O$ Mol.Wt.: 156.3 Handbook: d18)

	p.ol mass author	p.ol mass weighted	p.ol vol weighted
Pliska and Janicek 1965	5.42	8.22	9.03
Guadagni 1966	9.00	8.76	9.57
Randebrock 1971	9.61	7.76	8.57
Teranishi *et al.* 1974	8.30	8.21	9.02
Standard		8.23	9.05

Decane ($C_{10}H_{22}$ Mol.Wt.: 142.3 Handbook: d20)

	p.ol mass author	p.ol mass weighted	p.ol vol weighted
Laffort and Dravnieks 1973	4.95	5.36	6.13
Standard		5.36	6.13

Decanoic acid (capric acid) ($C_{10}H_{20}O_2$ Mol.Wt.: 172.3 Handbook: d42)

	p.ol mass author	p.ol mass weighted	p.ol vol weighted
Passy 1893*b*, 1893*c*	7.30	7.33	8.19
Backman 1917	7.07	7.07	7.93
Standard		7.20	8.06

1-Decanol ($C_{10}H_{22}O$ Mol.Wt.: 158.3 Handbook: d57)

	p.ol mass author	p.ol mass weighted	p.ol vol weighted
Backman 1917	6.88	6.88	7.70
Mullins 1955	4.71	6.46	7.28
Pliska 1962	5.58	7.56	8.38
Punter 1983	6.36	6.76	7.58
Standard		6.92	7.74

2-Decanone ($C_{10}H_{20}O$ Mol.Wt.: 156.3 Handbook: d62)

	p.ol mass author	p.ol mass weighted	p.ol vol weighted
Teranishi *et al.* 1974	7.23	7.14	7.95
Hall and Andersson 1983	6.96	7.43	8.24
Standard		7.28	8.10

trans-2-Decenal ($C_{10}H_{18}O$ Mol.Wt.: 154.3 Handbook: d66*)

	p.ol mass author	p.ol mass weighted	p.ol vol weighted
Teranishi *et al.* 1974	8.57	8.48	9.29
Hall and Andersson 1983	8.31	8.78	9.59
Standard		8.63	9.44

diacetone alcohol see: **4-Hydroxy-4-methylpentan-2-one**
diacetyl see: **2,3-Butanedione**
diallyl disulfide see: **2-Propenyldithio-2-propene**
diallyl sulfide see: **2-Propenylthio-2-propene**
diamyl sulfide see: **Pentylthiopentane**
dibenzyl sulfide see: **Benzylthiotoluene**

1,2-Dibromobenzene ($C_6H_4Br_2$ Mol.Wt.: 235.9 Handbook: b512)

	p.ol mass author	p.ol mass weighted	p.ol vol weighted
Backman 1917	7.55	7.55	8.54
Standard		7.55	8.54

Dibutylamine ($C_8H_{19}N$ Mol.Wt.: 129.3 Handbook: a820)

	p.ol mass author	p.ol mass weighted	p.ol vol weighted
Hellman and Small 1973, 1974	6.38	5.86	6.59
Laing *et al.* 1978	5.56	6.68	7.41
Standard		6.27	7.00

dibutyl sulfide see: **Butylthiobutane**

2,2-Dichloroacetic acid ($C_2H_2Cl_2O_2$ Mol.Wt.: 128.9 Handbook: a407)

	p.ol mass author	p.ol mass weighted	p.ol vol weighted
Backman 1917	6.64	6.64	7.37
Standard		6.64	7.37

3,4-Dichloroaniline ($C_6H_5Cl_2N$ Mol.Wt.: 162.0 Handbook: a1079)

	p.ol mass author	p.ol mass weighted	p.ol vol weighted
Punter 1983	6.54	6.94	7.77
Standard		6.94	7.77

1,2-Dichlorobenzene ($C_6H_4Cl_2$ Mol.Wt.: 147.0 Handbook: b535)

	p.ol mass author	p.ol mass weighted	p.ol vol weighted
Backman 1917	6.92	6.92	7.71
Punter 1983	5.37	5.77	6.56
Standard		6.35	7.14

1,4-Dichlorobenzene ($C_6H_4Cl_2$ Mol.Wt.: 147.0 Handbook: b537)

	p.ol mass author	p.ol mass weighted	p.ol vol weighted
Punter 1983	6.13	6.53	7.32
Standard		6.53	7.32

1,1'-dichlorodiethyl sulfide see: **1-Chloro-1-(chloroethylthio)ethane**

1,2-Dichloroethane (ethylene chloride) ($C_2H_4Cl_2$ Mol.Wt.: 99.0 Handbook: e216)

	p.ol mass author	p.ol mass weighted	p.ol vol weighted
Jones 1955*b*	2.82	4.33	4.95
Scherberger *et al.* 1958	3.08	4.50	5.12
May 1966	3.34	4.53	5.15
Dravnieks and O'Donnell 1971	3.72	4.19	4.81
Hellman and Small 1974	4.60	4.08	4.70
Standard		4.33	4.95

1,1-Dichloroethene ($C_2H_2Cl_2$ Mol.Wt.: 96.9 Handbook: e411)

	p.ol mass author	p.ol mass weighted	p.ol vol weighted
Janicek *et al.* 1960	2.26	3.84	4.45
Standard		3.84	4.45

trans-1,2-Dichloroethene ($C_2H_2Cl_2$ Mol.Wt.: 96.9 Handbook: e413)

	p.ol mass author	p.ol mass weighted	p.ol vol weighted
Lehmann and Schmidt–Kehl 1936	2.96	4.11	4.72
Standard		4.11	4.72

dichloroisopropyiether see: **1-Chloro-2-(1-chloroisopropoxy)propane**

Dichloromethane (methylene chloride) (CH_2Cl_2 Mol.Wt.: 84.9 Handbook: a252)

	p.ol mass author	p.ol mass weighted	p.ol vol weighted
Lehmann and Schmidt–Kehl 1936	2.96	4.11	4.66
Scherberger *et al.* 1958	2.81	4.23	4.78
May 1966	3.26	4.45	5.00
Leonardos *et al.* 1969	3.14	3.13	3.68
Basmadshijewa *et al.* 1970	4.93	4.08	4.63
Standard		4.00	4.55

2,4-Dichlorophenol ($C_6H_4Cl_2O$ Mol.Wt.: 163.0 Handbook: p631)

	p.ol mass author	p.ol mass weighted	p.ol vol weighted
Punter 1983	9.57	9.97	10.80
Standard		9.97	10.80

1,2-Dichloropropane (propylene chloride) ($C_3H_6Cl_2$ Mol.Wt.: 113.0 Handbook: p1161)

	p.ol mass author	p.ol mass weighted	p.ol vol weighted
Hellman and Small 1974	5.92	5.40	6.07
Standard		5.40	6.07

α-Dicyclopentadiene ($C_{10}H_{12}$ Mol.Wt.: 132.2 Handbook: d164)

	p.ol mass author	p.ol mass weighted	p.ol vol weighted
Kinkead *et al.* 1971*b*	7.80	6.87	7.61
Hellman and Small 1974	7.22	6.70	7.44
Standard		6.78	7.53

diethanolamine see: **3-Azapentane-1,5-diol**

Diethylamine ($C_4H_{11}N$ Mol.Wt.: 73.1 Handbook: a827)

	p.ol mass author	p.ol mass weighted	p.ol vol weighted
Geier 1936	7.50	7.08	7.56
Hellman and Small 1973, 1974	6.80	6.28	6.76
Cormack *et al.* 1974	7.05	6.68	7.16
Laing *et al.* 1978	4.37	5.49	5.97
Tkachev 1978	6.81	6.06	6.54
Anon 1980	7.05	6.08	6.56
Laing 1982	5.40	6.05	6.53
Standard		6.24	6.73

2-Diethylaminoethanol ($C_6H_{15}NO$ Mol.Wt.: 117.2 Handbook: e358)

	p.ol mass author	p.ol mass weighted	p.ol vol weighted
Hellman and Small 1973, 1974	7.30	6.78	7.47
Standard		6.78	7.47

Diethyl 2,3-dihydroxybutanedioate (diethyl tartrate)
($C_8H_{14}O_6$ Mol.Wt.: 206.2 Handbook: t17)

	p.ol mass author	p.ol mass weighted	p.ol vol weighted
Backman 1917	7.11	7.11	8.04
Standard		7.11	8.04

diethyldisulfide see: **Ethyldithioethane**
diethyl ketone see: **3-Pentanone**

Diethyl oxalate ($C_6H_{10}O_4$ Mol.Wt.: 146.1 Handbook: o266)

	p.ol mass author	p.ol mass weighted	p.ol vol weighted
Backman 1917	6.31	6.31	7.09
Standard		6.31	7.09

Diethyl selenide ($C_4H_{10}Se$ Mol.Wt.: 137.1 Handbook: s30)

	p.ol mass author	p.ol mass weighted	p.ol vol weighted
Katz and Talbert 1930	8.17	8.37	9.13
Standard		8.37	9.13

Diethyl succinate ($C_8H_{14}O_4$ Mol.Wt.: 174.2 Handbook: s193)

	p.ol mass author	p.ol mass weighted	p.ol vol weighted
Tempelaar 1913	6.10	6.69	7.55
Standard		6.69	7.55

diethyl sulfide see: **Ethylthioethane**

diethyl tartrate see: **Diethyl 2,3-dihydroxybutanedioate**

diethyl trisulfide see: **Ethyltrithioethane**

3,4-dihydro-2H-azole see: **1-Pyrroline**

3,4-Dihydro-2-ethoxy-1,2-pyran ($C_7H_{12}O_2$ Mol.Wt.: 128.2 Handbook: p1894*)

	p.ol mass author	p.ol mass weighted	p.ol vol weighted
Hellman and Small 1974	7.00	6.48	7.21
Standard		6.48	7.21

1,2-Dihydroxyethane diacetate (glycol diacetate) ($C_6H_{10}O_4$ Mol.Wt.: 146.1 Handbook: e285)

	p.ol mass author	p.ol mass weighted	p.ol vol weighted
Hellman and Small 1974	6.23	5.71	6.49
Standard		5.71	6.49

α,2-dihydroxytoluene see: **Salicyl alcohol**

diisoamyl sulfide see: **3-Methylbutylthio-3-methylbutane**

diisobutyl ketone see: **2,6-Dimethylheptan-4-one**

2,4-Diisocyanato-l-methylbenzene (tolylene diisocyanate)
($C_9H_6N_2O_2$ Mol.Wt.: 174.2 Handbook: t469*)

	p.ol mass author	p.ol mass weighted	p.ol vol weighted
Leonardos *et al.* 1969	4.82	4.81	5.67
Standard		4.81	5.67

diisopropanolamine see: **3-Aza-2,4-dimethylpentane-2,4-diol**

Diisopropylamine ($C_6H_{15}N$ Mol.Wt.: 101.2 Handbook: a848)

	p.ol mass author	p.ol mass weighted	p.ol vol weighted
Hellman and Small 1974	6.25	5.73	6.36
England *et al.* 1978	4.75	5.80	6.43
Standard		5.77	6.40

diisopropyl sulfide see: **Isopropylthioisopropane**

Di-2-methoxyphenyl carbonate (duotal)
($C_{15}H_{14}O_5$ Mol.Wt.: 274.3 Handbook: c166)

	p.ol mass author	p.ol mass weighted	p.ol vol weighted
Backman 1917	8.94	8.94	10.00
Standard		8.94	10.00

Dimethylacetamide (C_4H_9NO Mol.Wt.: 87.1 Handbook: a102)

	p.ol mass author	p.ol mass weighted	p.ol vol weighted
Leonardos *et al.* 1969	3.77	3.76	4.32
Standard		3.76	4.32

Dimethylamine (C_2H_7N Mol.Wt.: 45.1 Handbook: a850)

	p.ol mass author	p.ol mass weighted	p.ol vol weighted
Geier 1936	6.10	5.68	5.95
Leonardos *et al.* 1969	7.07	7.06	7.33
Stephens 1971	6.80	6.71	6.98
Tkachev 1978	7.53	6.78	7.05
Anon 1980	8.86	7.89	8.16
Standard		6.82	7.09

2-Dimethylaminoethanol ($C_4H_{11}NO$ Mol.Wt.: 89.1 Handbook: e359)

	p.ol mass author	p.ol mass weighted	p.ol vol weighted
Hellman and Small 1974	7.30	6.78	7.35
England *et al.* 1978	5.70	6.75	7.32
Standard		6.76	7.34

N,N-Dimethylaniline ($C_8H_{11}N$ Mol.Wt.: 121.2 Handbook: a1105)

	p.ol mass author	p.ol mass weighted	p.ol vol weighted
Backman 1917	6.05	6.05	6.75
Geier 1936	7.65	7.23	7.93
Deadman and Prigg 1959	7.92	7.60	8.30
Standard		6.96	7.66

2,5-Dimethylaniline ($C_8H_{11}N$ Mol.Wt.: 121.2 Handbook: b228)

	p.ol mass author	p.ol mass weighted	p.ol vol weighted
Backman 1917	6.57	6.57	7.27
Standard		6.57	7.27

1,2-Dimethylbenzene (*o*-xylene)
(C_8H_{10} Mol.Wt.: 106.2 Handbook: b660)

	p.ol mass author	p.ol mass weighted	p.ol vol weighted
Backman 1917	5.96	5.96	6.61
Backman 1918	6.09	5.61	6.26
Stuiver 1958	5.67	4.38	5.03
Naus 1962	6.00	5.38	6.03
Köster 1971	4.95	6.45	7.10
Anon 1980	6.11	5.14	5.79
Punter 1983	4.62	5.02	5.67
Standard		5.42	6.07

1,3-Dimethylbenzene (*m*-xylene)
(C_8H_{10} Mol.Wt.: 106.2 Handbook: b661)

	p.ol mass author	p.ol mass weighted	p.ol vol weighted
Backman 1917	5.92	5.92	6.57
Stuiver 1958	6.45	5.16	5.81
Gusev 1965	5.97	5.58	6.23
Dravnieks and O'Donnell 1971	5.88	6.35	7.00
Köster 1971	5.11	6.61	7.26
Anon 1980	6.28	5.31	5.96
Punter 1983	5.56	5.96	6.61
Standard		5.85	6.49

1,4-Dimethylbenzene (*p*-xylene)
(C_8H_{10} Mol.Wt.: 106.2 Handbook: b662)

	p.ol mass author	p.ol mass weighted	p.ol vol weighted
Backman 1917	5.83	5.83	6.48
Stuiver 1958	6.22	4.93	5.58
Leonardos *et al.* 1969	5.69	5.68	6.33
Köster 1971	5.09	6.59	7.24
Knuth 1973	6.09	5.86	6.51
Anon 1980	6.28	5.31	5.96
Punter 1983	5.04	5.44	6.09
Standard		5.67	6.31

3,3-Dimethylbutan-2-one (methyl tert.butyl ketone) ($C_6H_{12}O$ Mol.Wt.: 100.2 Handbook: b2921)

	p.ol mass author	p.ol mass weighted	p.ol vol weighted
Backman 1917	6.11	6.11	6.73
Standard		6.11	6.73

5,5-dimethyl-1,3-cyclohexanedione see: **Methone**
dimethyl disulfide see: **Methyldithiomethane**

Dimethylformamide (C_3H_7NO Mol.Wt.: 73.1 Handbook: f120)

	p.ol mass author	p.ol mass weighted	p.ol vol weighted
Leonardos *et al.* 1969	3.53	3.52	4.00
Standard		3.52	4.00

2,6-Dimethylheptan-4-ol ($C_9H_{20}O$ Mol.Wt.: 144.3 Handbook: h174)

	p.ol mass author	p.ol mass weighted	p.ol vol weighted
Hellman and Small 1974	6.72	6.20	6.98
Standard		6.20	6.98

2,6-Dimethylheptan-4-one (diisobutyl ketone) ($C_9H_{18}O$ Mol.Wt.: 142.2 Handbook: h187(1))

	p.ol mass author	p.ol mass weighted	p.ol vol weighted
Hellman and Small 1973, 1974	6.22	5.70	6.47
Standard		5.70	6.47

3,7-Dimethylocta-2,6-dienal (citral) ($C_{10}H_{16}O$ Mol.Wt.: 152.2 Handbook: c425)

	p.ol mass author	p.ol mass weighted	p.ol vol weighted
Passy 1892*a*, 1892*b*	6.65	6.80	7.60
Tempelaar 1913	7.21	7.80	8.60
Backman 1917	7.14	7.14	7.94
Ohma 1922	6.89	7.07	7.87
Appell 1969	9.30	6.91	7.71
Köster 1971	6.75	8.25	9.05
Standard		7.33	8.13

3,7-Dimethylocta-cis-2,6-dien-1-ol (nerol) ($C_{10}H_{18}O$ Mol.Wt.: 154.3 Handbook: n466)

	p.ol mass author	p.ol mass weighted	p.ol vol weighted
Schwarz 1955	8.82	8.88	9.69
Standard		8.88	9.69

3,7-Dimethylocta-1,6-dien-3-ol (linalol) ($C_{10}H_{18}O$ Mol.Wt.: 154.3 Handbook: 144)

	p.ol mass author	p.ol mass weighted	p.ol vol weighted
Appell 1969	8.85	6.46	7.27
Standard		6.46	7.27

3,7-Dimethyloct-6-enal (citronellal) ($C_{10}H_{18}O$ Mol.Wt.: 154.3 Handbook: c439)

	p.ol mass author	p.ol mass weighted	p.ol vol weighted
Appell 1969	9.52	7.13	7.94
Standard		7.13	7.94

3,7-Dimethyloct-6-en-1-ol (citronellol) ($C_{10}H_{20}O$ Mol.Wt.: 156.3 Handbook: c442)

	p.ol mass author	p.ol mass weighted	p.ol vol weighted
Tempelaar 1913	6.56	7.15	7.96
Appell 1969	9.92	7.53	8.34
Standard		7.34	8.15

2,4-Dimethylpentane (C_7H_{16} Mol.Wt.: 100.2 Handbook: p133)

	p.ol mass author	p.ol mass weighted	p.ol vol weighted
Dravnieks and Lafford 1972	3.06	3.16	3.78
Dravnieks 1974	2.52	3.72	4.34
Standard		3.44	4.06

2,3-Dimethylphenol ($C_8H_{10}O$ Mol.Wt.: 122.2 Handbook: b675)

	p.ol mass author	p.ol mass weighted	p.ol vol weighted
Stuiver 1958	9.30	8.01	8.72
Standard		8.01	8.72

2,4-Dimethylphenol ($C_8H_{10}O$ Mol.Wt.: 122.2 Handbook: b682)

	p.ol mass author	p.ol mass weighted	p.ol vol weighted
Backman 1917	9.00	9.00	9.71
Stuiver 1958	7.85	6.56	7.27
Standard		7.78	8.49

2,5-Dimethylphenol ($C_8H_{10}O$ Mol.Wt.: 122.2 Handbook: b680)

	p.ol mass author	p.ol mass weighted	p.ol vol weighted
Backman 1917	5.67	5.67	6.38
Stuiver 1958	9.30	8.01	8.72
Standard		6.84	7.55

2,6-Dimethylphenol ($C_8H_{10}O$ Mol.Wt.: 122.2 Handbook: b678)

	p.ol mass author	p.ol mass weighted	p.ol vol weighted
Stuiver 1958	9.70	8.41	9.12
Standard		8.41	9.12

3,4-Dimethylphenol ($C_8H_{10}O$ Mol.Wt.: 122.2 Handbook: b676)

	p.ol mass author	p.ol mass weighted	p.ol vol weighted
Stuiver 1958	8.52	7.23	7.94
Standard		7.23	7.94

3,5-Dimethylphenol ($C_8H_{10}O$ Mol.Wt.: 122.2 Handbook: b679)

	p.ol mass author	p.ol mass weighted	p.ol vol weighted
Stuiver 1958	10.39	9.10	9.81
Standard		9.10	9.81

dimethyl sulfide see: **Methylthiomethane**

dimethyl trisulfide see: **Methyltrithiomethane**

Dimethyl trithiocarbonate ($C_3H_6S_3$ Mol.Wt.: 138.3 Handbook: c183*)

	p.ol mass author	p.ol mass weighted	p.ol vol weighted
Katz and Talbert 1930	7.48	7.68	8.44
Standard		7.68	8.44

1,4-Dioxane ($C_4H_8O_2$ Mol.Wt.: 88.1 Handbook: d198)

	p.ol mass author	p.ol mass weighted	p.ol vol weighted
May 1966	3.20	4.39	4.96
Köster 1971	3.18	4.68	5.25
Hellman and Small 1973, 1974	5.53	5.01	5.58
Dravnieks and Laffort 1972	4.51	4.61	5.18
Dravnieks 1974	3.56	4.76	5.33
Standard		4.69	5.26

3,6-Dioxaoctan-1-ol (carbitol) ($C_6H_{14}O_3$ Mol.Wt.: 134.2 Handbook: d176)

	p.ol mass author	p.ol mass weighted	p.ol vol weighted
Hellman and Small 1973, 1974	5.92	5.40	6.15
Standard		5.40	6.15

3,6-Dioxaoctan-1-yl acetate (carbitol acetate) ($C_8H_{16}O_4$ Mol.Wt.: 176.2 Handbook: d177)

	p.ol mass author	p.ol mass weighted	p.ol vol weighted
Hellman and Small 1974	6.72	6.20	7.07
Standard		6.20	7.07

1,3-Dioxolane ($C_3H_6O_2$ Mol.Wt.: 74.1 Handbook: d208)

	p.ol mass author	p.ol mass weighted	p.ol vol weighted
Hellman and Small 1974	4.29	3.77	4.26
Standard		3.77	4.26

Diphenylamine ($C_{12}H_{11}N$ Mol.Wt.: 169.2 Handbook: a863)

	p.ol mass author	p.ol mass weighted	p.ol vol weighted
Backman 1917	6.80	6.80	7.65
Standard		6.80	7.65

diphenyl ether see: **Phenoxybenzene**

Diphenylmethane (benzylbenzene,ditan) ($C_{13}H_{12}$ Mol.Wt.: 168.2 Handbook: m271)

	p.ol mass author	p.ol mass weighted	p.ol vol weighted
Punter 1983	6.38	6.78	7.63
Standard		6.78	7.63

diphenyl sulfide see: **Phenylthiobenzene**

diphosgene see: **Trichloromethyl chloroformate**

Dipropylamine ($C_6H_{15}N$ Mol.Wt.: 101.2 Handbook: a893)

	p.ol mass author	p.ol mass weighted	p.ol vol weighted
Hellman and Small 1974	7.09	6.57	7.20
Standard		6.57	7.20

dipropyl disulfide see: **Propyldithiopropane**
dipropyl sulfide see: **Propylthiopropane**
ditan see: **Diphenylmethane**

Dodecane ($C_{12}H_{26}$ Mol.Wt.: 170.3 Handbook: d284)

	p.ol mass author	p.ol mass weighted	p.ol vol weighted
Lafford and Dravnieks 1973	4.43	4.84	5.69
Standard		4.84	5.69

1-Dodecanethiol ($C_{12}H_{26}S$ Mol.Wt.: 202.4 Handbook: d294)

	p.ol mass author	p.ol mass weighted	p.ol vol weighted
Kendall *et al.* 1968	9.09	8.68	9.61
Standard		8.68	9.61

Dodecanoic acid (lauric acid) ($C_{12}H_{24}O_2$ Mol.Wt.: 200.3 Handbook: d295)

	p.ol mass author	p.ol mass weighted	p.ol vol weighted
Passy 1893*b*, 1893*c*	7.00	7.03	7.95
Backman 1917	8.35	8.35	9.27
Standard		7.69	8.61

1-Dodecanol ($C_{12}H_{26}O$ Mol.Wt.: 186.3 Handbook: d311)

	p.ol mass author	p.ol mass weighted	p.ol vol weighted
Backman 1917	7.70	7.70	8.59
Mullins 1955	4.59	6.34	7.23
Pliska 1962	5.68	7.66	8.55
Punter 1983	5.92	6.32	7.21
Standard		7.00	7.90

tert.dodecyl mercaptan see: **2-Methylundecane-2-thiol**
duotal see: **Di-2-methoxyphenyl carbonate**
durene see: **1,2,4,5-Tetramethylbenzene**

1,4-Epoxybutane (tetrahydrofuran) (C_4H_8O Mol.Wt.: 72.1 Handbook: f220)

	p.ol mass author	p.ol mass weighted	p.ol vol weighted
May 1966	4.04	5.23	5.71
Kendall *et al.* 1968	4.04	4.65	5.13
Standard		4.94	5.42

1,4-Epoxyethane (ethylene oxide) (C_2H_4O Mol.Wt.: 44.1 Handbook: e241)

	p.ol mass author	p.ol mass weighted	p.ol vol weighted
Hellman and Small 1974	3.33	2.81	3.07
Standard		2.81	3.07

1,2-Epoxypropane (propylene oxide) (C_3H_6O Mol.Wt.: 58.1 Handbook: p1180)

	p.ol mass author	p.ol mass weighted	p.ol vol weighted
Hellman and Small 1974	4.62	4.10	4.48
Standard		4.10	4.48

Ethanal (acetaldehyde) (C_2H_4O Mol.Wt.: 44.1 Handbook: a18)

	p.ol mass author	p.ol mass weighted	p.ol vol weighted
Zwaardemaker 1914	6.16	6.08	6.34
Backman 1917	7.17	7.17	7.43
Katz and Talbert 1930	6.92	7.12	7.38
Pliska and Janicek 1965	2.75	5.55	5.81
Leonardos *et al.* 1969	6.42	6.41	6.67
Takhirov 1974	6.31	5.50	5.76
Teranishi *et al.* 1974	7.39	7.30	7.56
Anon 1980	8.57	7.60	7.86
Naus 1982	6.00	5.52	5.78
Standard		6.47	6.73

Ethanamide (acetamide) (C_2H_5NO Mol.Wt.: 59.1 Handbook: a60)

	p.ol mass author	p.ol mass weighted	p.ol vol weighted
Backman 1917	3.83	3.83	4.22
Standard		3.83	4.22

Ethane (C_2H_6 Mol.Wt.: 30.1 Handbook: e162)

	p.ol mass author	p.ol mass weighted	p.ol vol weighted
Mullins 1955	0.04	1.79	1.89
Laffort and Dravnieks 1973	1.60	2.01	2.11
Standard		1.90	2.00

1,2-Ethanediamine ($C_2H_8N_2$ Mol.Wt.: 60.1 Handbook: e207)

	p.ol mass author	p.ol mass weighted	p.ol vol weighted
Hellman and Small 1974	5.49	4.97	5.37
Standard		4.97	5.37

Ethane-1,2-dithiol ($C_2H_6S_2$ Mol.Wt.: 94.2 Handbook: e312)

	p.ol mass author	p.ol mass weighted	p.ol vol weighted
Katz and Talbert 1930	6.92	7.12	7.71
Standard		7.12	7.71

Ethanenitrile (acetonitrile) (C_2H_3N Mol.Wt.: 41.1 Handbook: a241)

	p.ol mass author	p.ol mass weighted	p.ol vol weighted
Dravnieks and Laffort 1972	3.55	3.65	3.88
Dravnieks 1974	2.71	3.91	4.14
Standard		3.78	4.01

Ethaneselenol (ethyl selenomercaptan) (C_2H_6Se Mol.Wt.: 109.0 Handbook: s30*)

	p.ol mass author	p.ol mass weighted	p.ol vol weighted
Katz and Talbert 1930	9.88	10.08	10.74
Standard		10.08	10.74

Ethanethiol (C_2H_6S Mol.Wt.: 62.1 Handbook: e328)

	p.ol mass author	p.ol mass weighted	p.ol vol weighted
Allison and Katz 1919	4.34	6.92	7.33
Katz and Talbert 1930	8.65	8.95	9.26
Thomas *et al.* 1943	8.30	8.40	8.81
Stuiver 1958	10.00	8.71	9.12
Sales 1958	8.48	8.54	8.95
Blinova 1965	7.88	8.45	8.86
Leonardos *et al.* 1969	8.61	8.60	9.01
Wilby 1969	9.00	8.92	9.33
Blanchard 1976	7.80	8.57	8.98
Bedborough and Trott 1979	9.48	8.78	9.19
Anon 1980	10.37	9.40	9.81
Standard		8.55	8.97

Ethanol (C_2H_6O Mol.Wt.: 46.1 Handbook: e336)

	p.ol mass author	p.ol mass weighted	p.ol vol weighted
Passy 1892*c*	3.61	3.76	4.04
Backman 1917	3.73	3.73	4.01
Grijns 1919	2.59	3.83	4.11
Henning 1924	3.74	2.93	3.21
Jung 1936	5.11	4.64	4.92
Mullins 1955	2.04	3.79	4.07
Scherberger *et al.* 1958	3.18	4.60	4.88
Janicek *et al.* 1960	3.06	4.64	4.92
Naus 1962	5.70	5.08	5.36
Pliska and Janicek 1965	1.12	3.92	4.20
Guadagni 1966	4.00	3.76	4.04
May 1966	4.03	5.22	5.50
Leonardos *et al.* 1969	4.72	4.71	4.99
Dravnieks and Laffort 1972	3.20	3.30	3.58
Dravnieks 1974	2.87	4.07	4.35
Anon 1980	6.20	5.23	5.51
Naus 1982	5.70	5.22	5.50
Standard		4.26	4.54

Ethanoyl chloride (acetyl chloride) (C_2H_3ClO Mol.Wt.: 78.5 Handbook: a183)

	p.ol mass author	p.ol mass weighted	p.ol vol weighted
Schley 1934	6.96	6.59	7.10
Standard		6.59	7.10

Ethene (ethylene) (C_2H_4 Mol.Wt.: 28.1 Handbook: e401)

	p.ol mass author	p.ol mass weighted	p.ol vol weighted
Mullins 1955	2.93	4.68	4.75
Deadman and Prigg 1959	3.90	3.58	3.65
Laffort and Dravnieks 1973	2.96	3.37	3.44
Hellman and Small 1974	3.51	2.99	3.06
Standard		3.66	3.73

ether see: **Ethoxyethane**

Ethoxyethane (ether) ($C_4H_{10}O$ Mol.Wt.: 74.1 Handbook: e477)

	p.ol mass author	p.ol mass weighted	p.ol vol weighted
Passy 1892*a*, 1892*b*	6.00	6.15	6.64
Allison and Katz 1919	2.23	4.81	5.30
Grijns 1919	4.30	5.54	6.03
Henning 1924	6.12	5.31	5.80
Jung 1936	4.46	3.99	4.48
Scherberger *et al.* 1958	3.68	5.10	5.59
Standard		5.15	5.64

2-Ethoxyethanol (ethyl glycol,cellosolve) ($C_4H_{10}O_2$ Mol.Wt.: 90.1 Handbook: e363)

	p.ol mass author	p.ol mass weighted	p.ol vol weighted
May 1966	4.05	5.24	5.81
Hellman and Small 1973, 1974	5.96	5.44	6.01
Standard		5.34	5.91

2-Ethoxyethyl acetate (cellosolve acetate) ($C_6H_{12}O_3$ Mol.Wt.: 132.2 Handbook: a446)

	p.ol mass author	p.ol mass weighted	p.ol vol weighted
Hellman and Small 1973, 1974	6.52	6.00	6.74
Standard		6.00	6.74

3-Ethoxy-4-hydroxybenzaldehyde (ethylvanillin) ($C_9H_{10}O_3$ Mol.Wt.: 166.2 Handbook: b118)

	p.ol mass author	p.ol mass weighted	p.ol vol weighted
Randebrock 1971	11.16	9.31	10.15
Standard		9.31	10.15

1-Ethoxynaphthalene ($C_{12}H_{12}O$ Mol.Wt.: 172.2 Handbook: n179)

	p.ol mass author	p.ol mass weighted	p.ol vol weighted
Backman 1917	9.09	9.09	9.95
Standard		9.09	9.95

2-Ethoxynaphthalene ($C_{12}H_{12}O$ Mol.Wt.: 172.2 Handbook: n180)

	p.ol mass author	p.ol mass weighted	p.ol vol weighted
Backman 1917	7.93	7.93	8.79
Standard		7.93	8.79

Ethyl acetate ($C_4H_8O_2$ Mol.Wt.: 88.1 Handbook: a199)

	p.ol mass author	p.ol mass weighted	p.ol vol weighted
Backman 1917	4.79	4.79	5.36
Allison and Katz 1919	3.16	5.74	6.31
Jung 1936	5.44	4.97	5.54
Jones 1955*b*	3.80	5.31	5.88
Gofmekler 1960	6.22	5.67	6.24
Jenicek *et al.* 1960	2.95	4.53	5.10
May 1966	3.74	4.93	5.50
Laffort and Dravnieks 1973	4.56	4.97	5.54
Hellman and Small 1973, 1974	4.63	4.11	4.68
Anon 1980	6.04	5.07	5.64
Standard		5.01	5.58

ethyl acetoacetate see: **Ethyl 3-oxobutanoate**

Ethyl acid succinate ($C_6H_{10}O_4$ Mol.Wt.: 146.1 Handbook: s223)

	p.ol mass author	p.ol mass weighted	p.ol vol weighted
Backman 1917	5.20	5.20	5.98
Standard		5.20	5.98

ethyl acrylate see: **Ethyl 2-propenoate**

Ethylamine (C_2H_7N Mol.Wt.: 45.1 Handbook: e163)

	p.ol mass author	p.ol mass weighted	p.ol vol weighted
Tkachev 1969	7.31	6.56	6.83
Hellman and Small 1974	6.31	5.79	6.06
Laing *et al.* 1978	5.19	6.31	6.58
Standard		6.22	6.49

N-Ethylaniline ($C_8H_{11}N$ Mol.Wt.: 121.2 Handbook: a1128)

	p.ol mass author	p.ol mass weighted	p.ol vol weighted
Backman 1917	5.74	5.74	6.44
Standard		5.74	6.44

Ethylbenzene (C_8H_{10} Mol.Wt.: 106.2 Handbook: b758)

	p.ol mass author	p.ol mass weighted	p.ol vol weighted
Köster 1971	6.39	7.89	8.54
Standard		7.89	8.54

Ethyl benzoate ($C_9H_{10}O_2$ Mol.Wt.: 150.2 Handbook: b1273)

	p.ol mass author	p.ol mass weighted	p.ol vol weighted
Jung 1936	7.15	6.68	7.48
Janicek *et al.* 1960	5.17	6.75	7.55
Appell 1969	9.22	6.83	7.63
Standard		6.76	7.55

Ethyl butanoate ($C_6H_{12}O_2$ Mol.Wt.: 116.2 Handbook: b2668)

	p.ol mass author	p.ol mass weighted	p.ol vol weighted
Tempelaar 1913	6.55	7.14	7.83
Backman 1917	6.76	6.76	7.45
Standard		6.95	7.64

2-Ethylbutan-1-ol ($C_6H_{14}O$ Mol.Wt.: 102.2 Handbook: b2890)

	p.ol mass author	p.ol mass weighted	p.ol vol weighted
Hellman and Small 1974	6.52	6.00	6.63
Standard		6.00	6.63

ethyl carbanilate see: **Ethyl n-phenylcarbamate**

Ethyl decanoate ($C_{12}H_{24}O_2$ Mol.Wt.: 200.3 Handbook: d47)

	p.ol mass author	p.ol mass weighted	p.ol vol weighted
Backman 1917	8.56	8.56	9.48
Schwarz 1955	8.89	8.95	9.87
Goldenberg 1967	8.92	7.69	8.61
Standard		8.40	9.32

Ethyldithioethane (diethyl disulfide) ($C_4H_{10}S_2$ Mol.Wt.: 122.3 Handbook: d238)

	p.ol mass author	p.ol mass weighted	p.ol vol weighted
Hermanides 1909	9.52	8.97	9.68
Zwaardemaker 1914	9.52	9.44	10.15
Wilby 1969	7.64	7.56	8.27
Standard		8.66	9.37

Ethyl dodecanoate ($C_{14}H_{28}O_2$ Mol.Wt.: 228.4 Handbook: d301)

	p.ol mass author	p.ol mass weighted	p.ol vol weighted
Backman 1917	8.70	8.70	9.68
Standard		8.70	9.68

ethylene see: **Ethene**
ethylene chloride see: **1,2-Dichloroethane**
ethylene oxide see: **1,2-Epoxyethane**

Ethyl 2,3-epoxy-3-phenylbutanoate (ethyl methyl phenyl glycidate) ($C_{12}H_{14}O_3$ Mol.Wt.: 206.2 Handbook: b2771)

	p.ol mass author	p.ol mass weighted	p.ol vol weighted
Appell 1969	12.16	9.77	10.70
Standard		9.77	10.70

Ethyl ethoxydithioformate (ethyl xanthate) ($C_3H_6OS_2$ Mol.Wt.: 122.2 Handbook: f168*)

	p.ol mass author	p.ol mass weighted	p.ol vol weighted
Deadman and Prigg 1959	7.52	7.20	7.91
Standard		7.20	7.91

Ethyl n-ethylcarbamate (ethylurethan) ($C_5H_{11}NO_2$ Mol.Wt.: 117.2 Handbook: c117)

	p.ol mass author	p.ol mass weighted	p.ol vol weighted
Backman 1917	5.98	5.98	6.67
Standard		5.98	6.67

Ethyl formate ($C_3H_6O_2$ Mol.Wt.: 74.1 Handbook: f135)

	p.ol mass author	p.ol mass weighted	p.ol vol weighted
Backman 1917	4.24	4.24	4.73
Standard		4.24	4.73

ethyl glycol see: **2-Ethoxyethanol**

Ethyl heptanoate ($C_9H_{18}O_2$ Mol.Wt.: 158.2 Handbook: h149)

	p.ol mass author	p.ol mass weighted	p.ol vol weighted
Backman 1917	6.74	6.74	7.56
Standard		6.74	7.56

Ethyl hexanoate ($C_8H_{16}O_2$ Mol.Wt.: 144.2 Handbook: h433)

	p.ol mass author	p.ol mass weighted	p.ol vol weighted
Backman 1917	8.34	8.34	9.12
Schwarz 1955	7.52	7.58	8.36
Standard		7.96	8.74

2-Ethylhexan-1-ol ($C_8H_{18}O$ Mol.Wt.: 130.2 Handbook: h491)

	p.ol mass author	p.ol mass weighted	p.ol vol weighted
Hellman and Small 1973, 1974	6.40	5.88	6.61
Standard		5.88	6.61

2-Ethylhexyl acetate ($C_{10}H_{20}O_2$ Mol.Wt.: 172.3 Handbook: a201)

	p.ol mass author	p.ol mass weighted	p.ol vol weighted
Hellman and Small 1974	6.15	5.63	6.49
Standard		5.63	6.49

2-Ethylhexyl 2-propenoate ($C_{11}H_{20}O_2$ Mol.Wt.: 184.3 Handbook: p1773*)

	p.ol mass author	p.ol mass weighted	p.ol vol weighted
Hellman and Small 1974	6.26	5.74	6.63
Standard		5.74	6.63

Ethyl 2-hydroxypropanoate (ethyl lactate) ($C_5H_{10}O_3$ Mol.Wt.: 118.1 Handbook: p1467)

	p.ol mass author	p.ol mass weighted	p.ol vol weighted
Backman 1917	5.10	5.10	5.79
Standard		5.10	5.79

Ethylidenenorbornene (C_9H_{12} Mol.Wt.: 120.2 Handbook: a22*)

	p.ol mass author	p.ol mass weighted	p.ol vol weighted
Kinkead *et al.* 1971*a*	7.31	6.38	7.08
Hellman and Small 1974	7.00	6.48	7.18
Standard		6.43	7.13

ethyl isobutyrate see: **Ethyl 2-methylpropanoate**

Ethyl isothiocyanate (C_3H_5NS Mol.Wt.: 87.1 Handbook: i228)

	p.ol mass author	p.ol mass weighted	p.ol vol weighted
Katz and Talbert 1930	5.21	5.41	5.97
Standard		5.41	5.97

ethyl isovalerate see; **Ethyl 3-methylbutanoate**

ethyl lactate see: **Ethyl 2-hydroxypropanoate**

Ethyl 3-methylbutanoate (ethyl isovalerate) ($C_7H_{14}O_2$ Mol.Wt.: 130.2 Handbook: b2806(1))

	p.ol mass author	p.ol mass weighted	p.ol vol weighted
Tempelaar 1913	5.34	5.93	6.66
Backman 1917	6.35	6.35	7.08
Standard		6.14	6.87

ethyl methyl phenyl glycidate see: **Ethyl 2,3-epoxy-3-phenylbutanoate**

Ethyl 2-methylpropanoate (ethyl isobutyrate) ($C_6H_{12}O_2$ Mol.Wt.: 116.2 Handbook: p1527)

	p.ol mass author	p.ol mass weighted	p.ol vol weighted
Backman 1917	6.02	6.02	6.71
Appell 1969	8.45	6.06	6.75
Standard		6.04	6.73

5-Ethyl-2-methylpyridine ($C_8H_{11}N$ Mol.Wt.: 121.2 Handbook: p2006)

	p.ol mass author	p.ol mass weighted	p.ol vol weighted
Kendall *et al.* 1968	7.89	7.48	8.18
Hellman and Small 1974	7.53	7.01	7.71
Standard		7.24	7.95

ethyl methyl sulfide see; **Methylthioethane**

N-Ethylmorpholine ($C_6H_{13}NO$ Mol.Wt.: 115.2 Handbook: m416)

	p.ol mass author	p.ol mass weighted	p.ol vol weighted
Hellman and Small 1974	6.40	5.88	6.56
Standard		5.88	6.56

Ethyl nonanoate ($C_{11}H_{22}O_2$ Mol.Wt.: 186.3 Handbook: n537)

	p.ol mass author	p.ol mass weighted	p.ol vol weighted
Backman 1917	8.00	8.00	8.89
Schwarz 1955	8.00	8.06	8.95
Standard		8.03	8.92

Ethyl octanoate ($C_{10}H_{20}O_2$ Mol.Wt.: 172.3 Handbook: o166)

	p.ol mass author	p.ol mass weighted	p.ol vol weighted
Backman 1917	8.70	8.70	9.56
Schwarz 1955	8.00	8.06	8.92
Standard		8.38	9.24

Ethyl 3-oxobutanoate (ethyl acetoacetate) ($C_6H_{10}O_3$ Mol.Wt.: 130.2 Handbook: b2828)

	p.ol mass author	p.ol mass weighted	p.ol vol weighted
Backman 1917	7.75	7.75	8.48
Standard		7.75	8.48

Ethyl pentanoate ($C_7H_{14}O_2$ Mol.Wt.: 130.2 Handbook: p222)

	p.ol mass author	p.ol mass weighted	p.ol vol weighted
Backman 1917	6.49	6.49	7.22
Stone 1963*b*, 1963*c*	6.45	7.23	7.96
Standard		6.86	7.59

Ethyl n-phenylcarbamate (phenylurethan,ethyl carbanilate) ($C_9H_{11}NO_2$ Mol.Wt.: 165.2 Handbook: c129)

	p.ol mass author	p.ol mass weighted	p.ol vol weighted
Backman 1917	7.36	7.36	8.20
Standard		7.36	8.20

Ethyl propanoate ($C_5H_{10}O_2$ Mol.Wt.: 102.1 Handbook: p1316)

	p.ol mass author	p.ol mass weighted	p.ol vol weighted
Backman 1917	6.41	6.41	7.04
Standard		6.41	7.04

Ethyl 2-propenoate ($C_5H_8O_2$ Mol.Wt.: 100.1 Handbook: p1773)

	p.ol mass author	p.ol mass weighted	p.ol vol weighted
Leonardos *et al.* 1969	8.72	8.71	9.33
Hellman and Small 1973, 1974	9.00	8.48	9.10
Anon 1980	9.09	8.12	8.74
Standard		8.44	9.06

Ethyl salicylate ($C_9H_{10}O_3$ Mol.Wt.: 166.2 Handbook: b1662)

	p.ol mass author	p.ol mass weighted	p.ol vol weighted
Backman 1917	6.92	6.92	7.76
Baldus 1936	8.46	7.36	8.20
Standard		7.14	7.98

ethyl selenomercaptan see: **Ethaneselenol**

Ethyl silicate ($C_8H_{20}O_4Si$ Mol.Wt.: 208.3 Handbook: o247)

	p.ol mass author	p.ol mass weighted	p.ol vol weighted
Hellman and Small 1974	4.51	3.99	4.93
Standard		3.99	4.93

Ethylthioethane (diethyl sulfide) ($C_4H_{10}S$ Mol.Wt.: 90.2 Handbook: s303)

	p.ol mass author	p.ol mass weighted	p.ol vol weighted
Allison and Katz 1919	4.92	7.50	8.08
Katz and Talbert 1930	8.00	8.20	8.78
Deadman and Prigg 1959	8.34	8.02	8.60
Wilby 1969	7.65	7.57	8.15
Anon 1980	8.85	7.88	8.46
Standard		7.84	8.41

Ethyl 2-thiolacetate (C_4H_8OS Mol.Wt.: 104.2 Handbook: a591)

	p.ol mass author	p.ol mass weighted	p.ol vol weighted
Deadman and Prigg 1959	8.09	7.77	8.41
Standard		7.77	8.41

Ethyltrithioethane (diethyl trisulfide) ($C_4H_{10}S_3$ Mol.Wt.: 154.3 Handbook: t746)

	p.ol mass author	p.ol mass weighted	p.ol vol weighted
Wilby 1969	8.27	8.19	9.00
Standard		8.19	9.00

Ethyl undecanoate ($C_{13}H_{26}O_2$ Mol.Wt.: 214.4 Handbook: h23)

	p.ol mass author	p.ol mass weighted	p.ol vol weighted
Schwarz 1955	8.30	8.36	9.31
Goldenberg 1967	9.52	8.29	9.24
Standard		8.32	9.28

ethylurethan see: **Ethyl n-ethylcarbamate**
ethylvanillin see: **3-Ethoxy-4-hydroxybenzaldehyde**
ethyl xanthate see: **Ethyl ethoxydithioformate**

Ethyne (acetylene) (C_2H_2 Mol.Wt.: 26.0 Handbook: e591)

	p.ol mass author	p.ol mass weighted	p.ol vol weighted
Deadman and Prigg 1959	3.62	3.30	3.34
Standard		3.30	3.34

eucalyptol see: **1,8-Cineole**
eugenol see: **4-Allyl-2-methoxyphenol**

Fenchone ($C_{10}H_{16}O$ Mol.Wt.: 152.2 Handbook: f11)

	p.ol mass author	p.ol mass weighted	p.ol vol weighted
Tempelaar 1913	5.64	6.23	7.03
Standard		6.23	7.03

Fluorotrichloromethane ($CFCl_3$ Mol.Wt.: 137.4 Handbook: m278*)

	p.ol mass author	p.ol mass weighted	p.ol vol weighted
Hellman and Small 1974	4.55	4.03	4.79
Standard		4.03	4.79

formaldehyde see: **Methanal**

Formic acid (CH_2O_2 Mol.Wt.: 46.0 Handbook: f115)

	p.ol mass author	p.ol mass weighted	p.ol vol weighted
Passy 1893*b*, 1893*c*	4.45	4.48	4.76
Zwaardemaker 1914	3.20	3.12	3.40
Backman 1917	4.65	4.65	4.93
Schley 1934	5.38	5.01	5.29
Guadagni 1966	3.35	3.11	3.39
Naus 1982	5.70	5.22	5.50
Standard		4.26	4.55

Furfural ($C_5H_4O_2$ Mol.Wt.: 96.1 Handbook: f277)

	p.ol mass author	p.ol mass weighted	p.ol vol weighted
Appell 1969	8.10	5.71	6.31
Makeicheva 1978	6.01	4.91	5.51
Bedborough and Trott 1979	6.60	5.90	6.50
Standard		5.50	6.11

geranial see: **Citral a**

glycol diacetate see: **1,2-Dihydroxyethane diacetate**

guaiacol see: **2-Methoxyphenol**

heliotropine see: **3,4-Methylenedioxybenzaldehyde**

trans-2,trans-4-Heptadienal ($C_7H_{10}O$ Mol.Wt.: 110.2 Handbook: h83*)

	p.ol mass author	p.ol mass weighted	p.ol vol weighted
Hall and Andersson 1983	7.25	7.72	8.38
Standard		7.72	8.38

Heptanal ($C_7H_{14}O$ Mol.Wt.: 114.2 Handbook: h90)

	p.ol mass author	p.ol mass weighted	p.ol vol weighted
Backman 1917	8.02	8.02	8.70
Deadman and Prigg 1959	8.22	7.90	8.58
Pliska and Janicek 1965	5.02	7.82	8.50
Teranishi *et al.* 1974	7.48	7.39	8.07
Hall and Andersson 1983	6.58	7.05	7.73
Standard		7.64	8.32

Heptane (C_7H_{16} Mol.Wt.: 100.2 Handbook: h93)

	p.ol mass author	p.ol mass weighted	p.ol vol weighted
Patty and Yant 1929	3.39	4.49	5.11
Mullins 1955	2.65	4.40	5.02
Jones 1955*b*	3.13	4.64	5.26
May 1966	3.03	4.22	4.84
Laffort and Dravnieks 1973	3.78	4.19	4.81
Standard		4.39	5.01

Heptanoic acid (oenanthic acid) ($C_7H_{14}O_2$ Mol.Wt.: 130.2 Handbook: h144)

	p.ol mass author	p.ol mass weighted	p.ol vol weighted
Passy 1893*b*, 1893*c*	6.53	6.56	7.29
Backman 1917	5.54	5.54	6.27
Hendriks 1979	7.49	7.16	7.89
Punter 1983	7.66	8.06	8.79
Standard		6.83	7.56

1-Heptanol ($C_7H_{16}O$ Mol.Wt.: 116.2 Handbook: h166)

	p.ol mass author	p.ol mass weighted	p.ol vol weighted
Backman 1917	6.34	6.84	7.53
Pliska 1962	4.76	6.74	7.43
Stone 1963*b*, 1963*c*	5.97	6.75	7.44
Stone *et al.* 1967	7.30	7.03	7.72
Corbit and Enger 1971	6.00	6.96	7.65
Laing 1975	5.62	7.00	7.69
Punter 1983	6.67	7.07	7.76
Standard		6.92	7.60

2-Heptanone ($C_7H_{14}O$ Mol.Wt.: 114.2 Handbook: h184)

	p.ol mass author	p.ol mass weighted	p.ol vol weighted
Teranishi *et al.* 1974	6.07	5.98	6.66
Hall and Andersson 1983	5.88	6.35	7.03
Standard		6.17	6.85

trans-2-Heptenal ($C_7H_{12}O$ Mol.Wt.: 112.2 Handbook: h197*)

	p.ol mass author	p.ol mass weighted	p.ol vol weighted
Teranishi *et al.* 1974	7.47	7.38	8.05
Eriksson *et al.* 1976	6.85	7.14	7.81
Hall and Andersson 1983	6.60	7.07	7.74
Standard		7.20	7.87

trans-3-Hepten-2-one ($C_7H_{12}O$ Mol.Wt.: 112.2 Handbook: h211)

	p.ol mass author	p.ol mass weighted	p.ol vol weighted
Teranishi *et al.* 1974	7.22	7.13	7.80
Standard		7.13	7.80

trans-2,trans-4-Hexadienal (C_6H_8O Mol.Wt.: 96.1 Handbook: h277)

	p.ol mass author	p.ol mass weighted	p.ol vol weighted
Teranishi *et al.* 1974	8.75	8.66	9.26
Standard		8.66	9.26

hexahydroazine see: **Piperidine**

Hexanal ($C_6H_{12}O$ Mol.Wt.: 100.2 Handbook: h316)

	p.ol mass author	p.ol mass weighted	p.ol vol weighted
Dravnieks and Laffort 1972	6.62	6.72	7.34
Teranishi *et al.* 1974	7.41	7.32	7.94
Eriksson *et al.* 1976	7.17	7.46	8.08
Logtenberg 1978	7.55	7.02	7.64
Hendriks 1979	7.40	7.07	7.69
Hall and Andersson 1983	7.37	7.84	8.46
Standard		7.24	7.86

Hexane (C_6H_{14} Mol.Wt.: 86.2 Handbook: h319)

	p.ol mass author	p.ol mass weighted	p.ol vol weighted
Patty and Yant 1929	3.05	4.15	4.71
Laffort and Dravnieks 1973	3.63	4.04	4.60
Standard		4.10	4.66

Hexanoic acid (caproic acid) ($C_6H_{12}O_2$ Mol.Wt.: 116.2 Handbook: h427)

	p.ol mass author	p.ol mass weighted	p.ol vol weighted
Passy 1893*b*, 1893*c*	7.39	7.42	8.11
Tempelaar 1913	6.28	6.87	7.56
Backman 1917	6.63	6.63	7.32
Hesse 1926	5.63	7.20	7.89
Hesse 1927	5.45	7.07	7.76
Punter 1983	7.69	8.09	8.78
Standard		7.22	7.90

1-Hexanol ($C_6H_{14}O$ Mol.Wt.: 102.2 Handbook: h478)

	p.ol mass author	p.ol mass weighted	p.ol vol weighted
Backman 1917	5.94	5.94	6.57
Mullins 1955	5.00	6.75	7.38
Pliska 1962	4.19	6.17	6.80
Cain 1969	5.46	7.01	7.64
Dravnieks and Laffort 1972	8.00	8.10	8.73
Stone *et al.* 1972	5.82	5.90	6.53
Dravnieks 1974	6.52	7.72	8.35
Hellman and Small 1974	7.40	6.88	7.51
Punter 1983	5.71	6.11	6.74
Standard		6.73	7.36

2-Hexanol ($C_6H_{14}O$ Mol.Wt.: 102.2 Handbook: h479)

	p.ol mass author	p.ol mass weighted	p.ol vol weighted
Dravnieks 1974	4.30	5.50	6.13
Standard		5.50	6.13

2-Hexanone ($C_6H_{12}O$ Mol.Wt.: 100.2 Handbook: h517)

	p.ol mass author	p.ol mass weighted	p.ol vol weighted
Backman 1917	6.51	6.51	7.13
Hall and Andersson 1983	5.33	5.80	6.42
Standard		6.15	6.78

trans-2-Hexenal ($C_6H_{10}O$ Mol.Wt.: 98.2 Handbook: h546)

	p.ol mass author	p.ol mass weighted	p.ol vol weighted
Teranishi *et al.* 1974	7.47	7.38	7.99
Eriksson *et al.* 1976	6.20	6.49	7.10
Hall and Andersson 1983	6.32	6.79	7.40
Standard		6.88	7.50

cis-3-Hexenal ($C_6H_{10}O$ Mol.Wt.: 98.2 Handbook: h546*)

	p.ol mass author	p.ol mass weighted	p.ol vol weighted
Teranishi *et al.* 1974	8.66	8.57	9.18
Standard		8.57	9.18

trans-4-Hexenal ($C_6H_{10}O$ Mol.Wt.: 98.2 Handbook: h546*)

	p.ol mass author	p.ol mass weighted	p.ol vol weighted
Teranishi *et al.* 1974	8.59	8.50	9.11
Standard		8.50	9.11

Hexyl acetate ($C_8H_{16}O$ Mol.Wt.: 144.2 Handbook: a208)

	p.ol mass author	p.ol mass weighted	p.ol vol weighted
Stone *et al.* 1972	5.64	5.72	6.50
Standard		5.72	6.50

Hydrocyanic acid (CHN Mol.Wt.: 27.0 Handbook: h724)

	p.ol mass author	p.ol mass weighted	p.ol vol weighted
Smolczyk and Cobler 1930	5.96	6.17	6.22
Standard		6.17	6.22

Hydrogen chloride (HCl Mol.Wt.: 36.5 Handbook: h50)

	p.ol mass author	p.ol mass weighted	p.ol vol weighted
Schley 1934	5.35	4.98	5.16
Leonardos *et al.* 1969	4.83	4.82	5.00
Takhirov 1974	6.42	5.61	5.79
Naus 1982	5.16	4.68	4.86
Standard		5.02	5.20

Hydrogen sulfide (H_2S Mol.Wt.: 34.1 Handbook: I, h68)

	p.ol mass author	p.ol mass weighted	p.ol vol weighted
Valentin 1848, 1850	5.70	6.84	6.99
Henning 1924	10.00	9.19	9.34
Katz and Talbert 1930	6.75	6.95	7.10
Thomas *et al.* 1943	7.46	7.56	7.71
Baikov 1963	7.69	7.53	7.68
Basmadzhieva and Argirova 1968	7.92	7.07	7.22
Leonardos *et al.* 1969	8.68	8.67	8.82
Wilby 1969	8.20	8.12	8.27
Stephens 1971	7.18	7.09	7.24
Randebrock 1971	7.92	6.07	6.22
Nishida *et al.* 1975	8.06	8.21	8.36
Logtenberg 1978	8.70	8.17	8.32
Bedborough and Trott 1979	8.45	7.75	7.90
Anon 1980	9.16	8.19	8.34
Naus 1982	7.00	6.52	6.67
Standard		7.59	7.75

2-Hydroxybenzaldehyde (salicyaldehyde) ($C_7H_6O_2$ Mol.Wt.: 122.1 Handbook: b122)

	p.ol mass author	p.ol mass weighted	p.ol vol weighted
Tempelaar 1913	7.85	8.44	9.15
Backman 1917	6.40	6.40	7.11
Standard		7.42	8.13

3-Hydroxybutanoic acid ($C_4H_8O_3$ Mol.Wt.: 104.1 Handbook: b2783)

	p.ol mass author	p.ol mass weighted	p.ol vol weighted
Backman 1917	4.37	4.37	5.01
Standard		4.37	5.01

4-Hydroxy-3-methoxybenzaldehyde (vanillin) ($C_8H_8O_3$ Mol.Wt.: 152.2 Handbook: b139)

	p.ol mass author	p.ol mass weighted	p.ol vol weighted
Passy 1892*a*, 1892*b*	8.80	8.95	9.75
Tempelaar 1913	9.75	10.34	11.14
Backman 1917	8.76	8.76	9.56
Baldus 1936	12.00	10.90	11.70
Appell 1969	12.00	9.61	10.41
Randebrock 1971	11.22	9.37	10.17
Standard		9.65	10.46

4-Hydroxy-4-methylpentan-2-one (diacetone alcohol) ($C_6H_{12}O_2$ Mol.Wt.: 116.2 Handbook: p362)

	p.ol mass author	p.ol mass weighted	p.ol vol weighted
Hellman and Small 1974	5.88	5.36	6.05
Standard		5.36	6.05

1-Hydroxynaphthalene (1-naphthol) ($C_{10}H_8O$ Mol.Wt.: 144.2 Handbook: n190)

	p.ol mass author	p.ol mass weighted	p.ol vol weighted
Backman 1917	8.40	8.40	9.18
Standard		8.40	9.18

2-Hydroxynaphthalene (2-naphthol) ($C_{10}H_8O$ Mol.Wt.: 144.2 Handbook: n191)

	p.ol mass author	p.ol mass weighted	p.ol vol weighted
Backman 1917	6.58	6.58	7.36
Standard		6.58	7.36

4-Hydroxynonanoic acid lactone ($C_9H_{16}O_2$ Mol.Wt.: 156.2 Handbook: n537*)

	p.ol mass author	p.ol mass weighted	p.ol vol weighted
Appell 1969	12.10	9.71	10.52
Standard		9.71	10.52

15-Hydroxypentadecanoic acid lactone ($C_{15}H_{28}O_2$ Mol.Wt.: 240.4 Handbook: p37*)

	p.ol mass author	p.ol mass weighted	p.ol vol weighted
Amoore 1977	6.75	5.99	6.99
Standard		5.99	6.99

4-Hydroxyundecanoic acid lactone ($C_{11}H_{20}O_2$ Mol.Wt.: 184.3 Handbook: h30)

	p.ol mass author	p.ol mass weighted	p.ol vol weighted
Appell 1969	12.22	9.83	10.72
Standard		9.83	10.72

Indene (C_9H_8 Mol.Wt.: 116.2 Handbook: i49)

	p.ol mass author	p.ol mass weighted	p.ol vol weighted
Deadman and Prigg 1959	7.69	7.37	8.06
Standard		7.37	8.06

Indole (C_8H_7N Mol.Wt.: 117.2 Handbook: i65)

	p.ol mass author	p.ol mass weighted	p.ol vol weighted
Tempelaar 1913	9.22	9.81	10.50
Standard		9.81	10.50

Iodobenzene (C_6H_5I Mol.Wt.: 204.0 Handbook: b828)

	p.ol mass author	p.ol mass weighted	p.ol vol weighted
Backman 1917	7.53	7.53	8.46
Standard		7.53	8.46

1-Iodobutane (C_4H_9I Mol.Wt.: 184.0 Handbook: b2568)

	p.ol mass author	p.ol mass weighted	p.ol vol weighted
Dravnieks and Laffort 1972	5.34	5.44	6.32
Dravnieks 1974	4.23	5.43	6.31
Standard		5.43	6.32

Iodoethane (C_2H_5I Mol.Wt.: 156.0 Handbook: e253)

	p.ol mass author	p.ol mass weighted	p.ol vol weighted
Backman 1917	5.07	5.07	5.88
Standard		5.07	5.88

iodoform see: **Triiodomethane**

l-Iodo-2-methylbenzene (C_7H_7I Mol.Wt.: 218.0 Handbook: t538)

	p.ol mass author	p.ol mass weighted	p.ol vol weighted
Backman 1917	8.04	8.04	9.00
Standard		8.04	9.00

2-Iodophenol (C_6H_5IO Mol.Wt.: 220.0 Handbook: p656)

	p.ol mass author	p.ol mass weighted	p.ol vol weighted
Kendall *et al.* 1968	11.05	10.64	11.60
Standard		10.64	11.60

α-ionone see: **4-(2,6,6-Trimethyl-2-cyclohexen-1-yl)-3-buten-2-one**
β-ionone see: **4-(2,6,6-Trimethyl-1-cyclohexen-1-yl)-3-buten-2-one**
isoamyl acetate see: **Isopentyl acetate**
isoamyl alcohol see: **3-Methylbutan-1-ol**
isoamyl butanoate see: **Isopentyl butanoate**
isoamyl isovalerate see: **Isopentyl 3-methylbutanoate**
isoamyl mercaptan see: **3-Methylbutane-1-thiol**
isoamyl propanoate see: **Isopentyl propanoate**
isobutane see: **2-Methylpropane**

2-Isobutoxyethanol (isobutyl cellosolve) ($C_6H_{14}O_2$ Mol.Wt.: 118.2 Handbook: e368(1))

	p.ol mass author	p.ol mass weighted	p.ol vol weighted
Hellman and Small 1974	7.05	6.53	7.22
Standard		6.53	7.22

Isobutyl acetate ($C_6H_{12}O_2$ Mol.Wt.: 116.2 Handbook: a215)

	p.ol mass author	p.ol mass weighted	p.ol vol weighted
Backman 1917	5.69	5.69	6.38
May 1966	4.76	5.95	6.64
Hellman and Small 1974	5.76	5.24	5.93
Standard		5.63	6.32

isobutyl acrylate see: **Isobutyl 2-propenoate**
isobutyl alcohol see: **2-Methylpropan-1-ol**

Isobutyl butanoate ($C_8H_{16}O_2$ Mol.Wt.: 144.2 Handbook: b2672)

	p.ol mass author	p.ol mass weighted	p.ol vol weighted
Backman 1917	7.11	7.11	7.89
Laing 1975	4.74	6.12	6.90
Standard		6.62	7.40

isobutyl cellosolve see: **2-Isobutoxyethanol**

isobutylene see: **2-Methylpropene**

isobutyl isobutyrate see: **Isobutyl 2-methylpropanoate**

isobutyl isovalerate see: **Isobutyl 3-methylbutanoate**

isobutyl mercaptan see: **2-Methylpropane-1-thiol**

Isobutyl 3-methylbutanoate (isobutyl isovalerate) ($C_9H_{18}O_2$ Mol.Wt.: 158.2 Handbook: b2807)

	p.ol mass author	p.ol mass weighted	p.ol vol weighted
Backman 1917	5.80	5.80	6.62
Standard		5.80	6.62

Isobutyl 2-methylpropanoate (isobutyl isobutyrate) ($C_8H_{16}O_2$ Mol.Wt.: 144.2 Handbook: p1530)

	p.ol mass author	p.ol mass weighted	p.ol vol weighted
Punter 1983	5.43	5.83	6.61
Standard		5.83	6.61

Isobutyl propanoate ($C_7H_{14}O_2$ Mol.Wt.: 130.2 Handbook: p1322)

	p.ol mass author	p.ol mass weighted	p.ol vol weighted
Backman 1917	6.39	6.39	7.12
Standard		6.39	7.12

Isobutyl 2-propenoate (isobutyl acrylate) ($C_7H_{12}O_2$ Mol.Wt.: 128.2 Handbook: p1774)

	p.ol mass author	p.ol mass weighted	p.ol vol weighted
Hellman and Small 1974	8.00	7.48	8.21
Standard		7.48	8.21

isobutyraldehyde see: **2-Methylpropanal**

isobutyric acid see: **2-Methylpropanoic acid**

isocaproic acid see: **4-Methylpentanoic acid**

Isopentyl acetate (isoamyl acetate) ($C_7H_{14}O_2$ Mol.Wt.: 130.2 Handbook: a223)

	p.ol mass author	p.ol mass weighted	p.ol vol weighted
Hermanides 1909	7.05	6.50	7.23
Zwaardemaker 1914	7.05	6.97	7.70
Backman 1917	6.65	6.65	7.38
Katz and Talbert 1930	7.75	7.95	8.68
Jung 1936	6.70	6.23	6.96
Kerka and Humphreys 1956	6.70	6.32	7.05
Pliska and Janicek	5.31	8.11	8.84
Appell 1969	8.40	6.01	6.74
Punter 1983	7.12	7.52	8.25
Standard		6.91	7.65

Isopentyl butanoate (isoamyl butanoate) ($C_9H_{18}O_2$ Mol.Wt.: 158.2 Handbook: b2678)

	p.ol mass author	p.ol mass weighted	p.ol vol weighted
Backman 1917	6.87	6.87	7.69
Standard		6.87	7.69

Isopentyl 3-methylbutanoate (isoamyl isovalerate) ($C_{10}H_{20}O_2$ Mol.Wt.: 172.3 Handbook: b2806*)

	p.ol mass author	p.ol mass weighted	p.ol vol weighted
Backman 1917	6.19	6.19	7.05
Katz and Talbert 1930	7.33	7.53	8.39
Standard		6.86	7.72

Isopentyl propanoate (isoamyl propanoate) ($C_8H_{16}O_2$ Mol.Wt.: 144.2 Handbook: p1325)

	p.ol mass author	p.ol mass weighted	p.ol vol weighted
Backman 1917	6.70	6.70	7.48
Standard		6.70	7.48

isophorone see: **3,5,5-Trimethyl-2-cyclohexen-1-one**

isopral see: **1,1,1-Trichloropropan-2-ol**

Isopropenylbenzene (α-methylstyrene) (C_9H_{10} Mol.Wt.: 118.2 Handbook: p1754)

	p.ol mass author	p.ol mass weighted	p.ol vol weighted
Wolf *et al.* 1956	3.97	6.17	6.86
Hellman and Small 1974	6.60	6.08	6.77
Standard		6.12	6.82

2-Isopropoxypropane (isopropyl ether) ($C_6H_{14}O$ Mol.Wt.: 102.2 Handbook: e497)

	p.ol mass author	p.ol mass weighted	p.ol vol weighted
Hellman and Small 1974	7.15	6.63	7.26
Standard		6.63	7.26

Isopropyl acetate ($C_5H_{10}O_2$ Mol.Wt.: 102.1 Handbook: a217)

	p.ol mass author	p.ol mass weighted	p.ol vol weighted
Backman 1917	4.52	4.52	5.15
Jung 1936	5.72	5.25	5.88
May 1966	3.85	5.04	5.67
Hellman and Small 1974	5.68	5.16	5.79
Standard		4.99	5.62

Isopropylamine (C_3H_9N Mol.Wt.: 59.1 Handbook: a904)

	p.ol mass author	p.ol mass weighted	p.ol vol weighted
Hellman and Small 1974	6.30	5.78	6.17
Standard		5.78	6.17

isopropyl alcohol see: **2-Propanol**

Isopropylbenzene (cumene) (C_9H_{12} Mol.Wt.: 120.2 Handbook: b847)

	p.ol mass author	p.ol mass weighted	p.ol vol weighted
Elfimova 1966	7.60	6.89	7.59
Köster 1971	6.60	8.10	8.80
Hellman and Small 1974	7.40	6.88	7.58
Anon 1980	7.13	6.16	6.86
Punter 1983	6.19	6.59	7.29
Standard		6.92	7.62

isopropyl ether see: **2-Isopropoxypropane**
isopropyl mercaptan see: **2-Propanethiol**

1-Isopropyl-2-methylbenzene (o-cymene) ($C_{10}H_{14}O$ Mol.Wt.: 134.2 Handbook: b848)

	p.ol mass author	p.ol mass weighted	p.ol vol weighted
Backman 1917	8.35	8.35	9.10
Standard		8.35	9.10

1-Isopropyl-4-methylbenzene (p-cymene) ($C_{10}H_{14}$ Mol.Wt.: 134.2 Handbook: b850)

	p.ol mass author	p.ol mass weighted	p.ol vol weighted
Backman 1917	7.92	7.92	8.67
Standard		7.92	8.67

2-Isopropyl-5-methylphenol (thymol) ($C_{10}H_{14}O$ Mol.Wt.: 150.2 Handbook: b797)

	p.ol mass author	p.ol mass weighted	p.ol vol weighted
Tempelaar 1913	7.42	8.01	8.81
Backman 1917	7.19	7.19	7.99
Van Anrooij 1931	7.39	6.65	7.45
Baldus 1936	7.30	6.20	7.00
Standard		7.02	7.81

5-Isopropyl-2-methylphenol (carvacrol) ($C_{10}H_{14}O$ Mol.Wt.: 150.2 Handbook: b800)

	p.ol mass author	p.ol mass weighted	p.ol vol weighted
Tempelaar 1913	7.00	7.59	8.39
Backman 1917	7.86	7.86	8.66
Van Anrooij 1931	6.74	6.00	6.80
Standard		7.15	7.95

2-Isopropylthiopropane (diisopropyl sulfide) ($C_6H_{14}S$ Mol.Wt.: 118.2 Handbook: s312)

	p.ol mass author	p.ol mass weighted	p.ol vol weighted
Wilby 1969	7.75	7.67	8.36
Standard		7.67	8.36

isovaleraldehyde see: **3-Methylbutanal**
isovaleric acid see: **3-Methylbutanoic acid**
lauric acid see: **Dodecanoic acid**

Limonene ($C_{10}H_{16}$ Mol.Wt.: 136.2 Handbook: 142)

	p.ol mass author	p.ol mass weighted	p.ol vol weighted
Appell 1969	8.00	5.61	6.36
Standard		5.61	6.36

linalool see: **3,7-Dimethylocta-1,6-dien-3-ol**

Linalyl acetate ($C_{12}H_{20}O_2$ Mol.Wt.: 196.3 Handbook: 145)

	p.ol mass author	p.ol mass weighted	p.ol vol weighted
Tempelaar 1913	7.03	7.62	8.53
Van Anrooij 1931	7.40	6.66	7.57
Standard		7.14	8.05

Menthol ($C_{10}H_{20}O$ Mol.Wt.: 156.3 Handbook: m164)

	p.ol mass author	p.ol mass weighted	p.ol vol weighted
Tempelaar 1913	4.17	4.76	5.57
Backman 1917	6.66	6.66	7.47
Baldus 1936	9.40	8.30	9.11
Standard		6.57	7.38

mesitylene see: **1,3,5-Trimethylbenzene**
mesityl oxide see: **4-Methylpent-3-en-2-one**
mesotan see: **2-Oxapropyl salicylate**

Methanal (formaldehyde) (CH_2O Mol.Wt.: 30.0 Handbook: f106)

	p.ol mass author	p.ol mass weighted	p.ol vol weighted
Backman 1917	7.46	7.46	7.56
Buchberg *et al.* 1961	5.81	5.93	6.03
Pliska and Janicek 1969	1.92	4.72	4.82
Leonardos *et al.* 1969	5.92	5.91	6.01
Takhirov 1974	7.18	6.37	6.47
Makeicheva 1978	7.11	6.01	6.11
Anon 1980	6.31	5.34	5.44
Standard		5.97	6.06

Methanethiol (CH_4S Mol.Wt.: 48.1 Handbook: m345)

	p.ol mass author	p.ol mass weighted	p.ol vol weighted
Katz and Talbert 1930	7.09	7.29	7.59
Guadagni *et al.* 1966	9.70	9.46	9.76
Leonardos *et al.* 1969	8.38	8.37	8.67
Wilby 1969	8.70	8.62	8.92
Blanchard 1976	8.53	9.30	9.60
Bedborough and Trott 1979	9.80	9.10	9.40
Anon 1980	9.62	8.65	8.95
Standard		8.68	8.98

Methanol (CH_4O Mol.Wt.: 32.0 Handbook: m349)

	p.ol mass author	p.ol mass weighted	p.ol vol weighted
Passy 1892*c*	3.00	3.15	3.28
Zwaardemaker 1914	3.22	3.14	3.27
Backman 1917	3.02	3.02	3.15
Grijns 1919	2.66	3.90	4.03
Jung 1936	4.44	3.97	4.10
Mullins 1955	1.71	3.46	3.59
Scherberger *et al.* 1958	2.71	4.13	4.26
Janicek *et al.* 1960	2.39	3.97	4.10
Pliska and Janicek 1965	0.58	3.38	3.51
May 1966	2.10	3.29	3.42
Leonardos *et al.* 1969	3.88	3.87	4.00
Hellman and Small 1974	5.26	4.74	4.87
Anon 1980	4.13	3.16	3.29
Naus 1982	5.39	4.91	5.04
Standard		3.73	3.85

Methone (5,5-dimethyl-1,3-cyclohexanedione) ($C_8H_{12}O_2$ Mol.Wt.: 140.2 Handbook: c735)

	p.ol mass author	p.ol mass weighted	p.ol vol weighted
Stone *et al.* 1967	4.36	4.09	4.86
Standard		4.09	4.86

4-Methoxybenzaldehyde (anisaldehyde) ($C_8H_8O_2$ Mol.Wt.: 136.2 Handbook: b154)

	p.ol mass author	p.ol mass weighted	p.ol vol weighted
Van Anrooij 1931	7.47	6.73	7.48
Standard		6.73	7.48

Methoxybenzene (anisole) (C_7H_8O Mol.Wt.: 108.2 Handbook: b859)

	p.ol mass author	p.ol mass weighted	p.ol vol weighted
Hendriks 1979	6.26	5.93	6.58
Punter 1983	5.83	6.23	6.88
Standard		6.08	6.73

2-Methoxyethanol (methyl glycol, methyl cellosolve) ($C_3H_8O_2$ Mol.Wt.: 76.1 Handbook: e373)

	p.ol mass author	p.ol mass weighted	p.ol vol weighted
May 1966	3.72	4.91	5.41
Hellman and Small 1973, 1974	6.52	6.00	6.50
Standard		5.45	5.96

2-Methoxyethyl acetate (methyl cellosolve acetate) ($C_5H_{10}O_3$ Mol.Wt.: 118.1 Handbook: a219)

	p.ol mass author	p.ol mass weighted	p.ol vol weighted
Hellman and Small 1973, 1974	5.80	5.28	5.97
Standard		5.28	5.97

1-Methoxy-4-methylbenzene (methyl p-cresol) ($C_8H_{10}O$ Mol.Wt.: 122.2 Handbook: t549)

	p.ol mass author	p.ol mass weighted	p.ol vol weighted
Appell 1969	9.70	7.31	8.02
Standard		7.31	8.02

methoxymethyl salicylate see: **2-Oxapropyl salicylate**

1-Methoxynaphthalene ($C_{11}H_{10}O$ Mol.Wt.: 158.2 Handbook: n215)

	p.ol mass author	p.ol mass weighted	p.ol vol weighted
Backman 1917	9.07	9.07	9.89
Standard		9.07	9.89

2-Methoxynaphthalene ($C_{11}H_{10}O$ Mol.Wt.: 158.2 Handbook: n216)

	p.ol mass author	p.ol mass weighted	p.ol vol weighted
Passy 1892*a*	8.65	8.80	9.62
Backman 1917	7.95	7.95	8.77
Standard		8.38	9.20

2-Methoxyphenol (guaiacol) ($C_7H_8O_2$ Mol.Wt.: 124.2 Handbook: p662)

	p.ol mass author	p.ol mass weighted	p.ol vol weighted
Hermanides 1909	8.44	7.89	8.60
Zwaardemaker 1914	8.44	8.36	9.07
Backman 1917	8.61	8.61	9.32
Standard		8.28	9.00

1-Methoxy-4-propenylbenzene (anethole) ($C_{10}H_{12}O$ Mol.Wt.: 148.2 Handbook: b864)

	p.ol mass author	p.ol mass weighted	p.ol vol weighted
Tempelaar 1913	6.85	7.44	8.23
Rocen 1920	6.72	7.27	8.06
Standard		7.35	8.15

N-Methylacetanilide ($C_9H_{11}NO$ Mol.Wt.: 149.2 Handbook: a152)

	p.ol mass author	p.ol mass weighted	p.ol vol weighted
Backman 1917	6.34	6.34	7.13
Standard		6.34	7.13

Methyl acetate ($C_3H_6O_2$ Mol.Wt.: 74.1 Handbook: a221)

	p.ol mass author	p.ol mass weighted	p.ol vol weighted
Zwaardemaker 1914	5.70	5.62	6.11
Backman 1917	4.17	4.17	4.66
Gofmekler 1960	6.30	5.75	6.24
Janicek *et al.* 1960	2.28	3.86	4.35
Naus 1962	6.15	5.53	6.02
May 1966	3.26	4.45	4.94
Anon 1980	4.66	3.69	4.18
Standard		4.72	5.21

methyl acetophenone see: **1-Acetyl-4-methylbenzene**
methyl acrylate see: **Methyl 2-propenoate**

Methylamine (CH_5N Mol.Wt.: 31.1 Handbook: m176)

	p.ol mass author	p.ol mass weighted	p.ol vol weighted
Leonardos *et al.* 1969	7.57	7.56	7.67
Nishida *et al.* 1975	7.19	7.34	7.45
Anon 1980	8.92	7.95	8.06
Standard		7.61	7.73

Methyl 2-aminobenzoate (methyl anthranylate) ($C_8H_9NO_2$ Mol.Wt.: 151.2 Handbook: b1319)

	p.ol mass author	p.ol mass weighted	p.ol vol weighted
Zwaardemaker 1914	8.22	8.14	8.94
Katz and Talbert 1930	7.24	7.44	8.24
Appell 1969	11.22	8.83	9.63
Standard		8.14	8.94

2-Methylaminoethanol (C_3H_9NO Mol.Wt.: 75.1 Handbook: e377)

	p.ol mass author	p.ol mass weighted	p.ol vol weighted
Hellman and Small 1974	5.50	4.98	5.48
Standard		4.98	5.48

Methyl 3-amino-4-hydroxybenzoate ($C_8H_9NO_3$ Mol.Wt.: 167.2 Handbook: b1360)

	p.ol mass author	p.ol mass weighted	p.ol vol weighted
Backman 1917	6.21	6.21	7.05
Standard		6.21	7.05

N-Methylaniline (C_7H_9N Mol.Wt.: 107.2 Handbook: a1157)

	p.ol mass author	p.ol mass weighted	p.ol vol weighted
Backman 1917	5.11	5.11	5.76
Standard		5.11	5.76

2-Methylaniline (*o*-toluidine) (C_7H_9N Mol.Wt.: 107.2 Handbook: t283)

	p.ol mass author	p.ol mass weighted	p.ol vol weighted
Backman 1917	5.33	5.33	5.98
Huijer 1918	4.54	5.60	6.25
Stuiver 1958	6.96	5.67	6.32
Standard		5.53	6.18

3-Methylaniline (*m*-toluidine) (C_7H_9N Mol.Wt.: 107.2 Handbook: t286)

	p.ol mass author	p.ol mass weighted	p.ol vol weighted
Backman 1917	5.47	5.47	6.12
Huijer 1918	4.59	5.65	6.30
Stuiver 1958	5.70	4.41	5.06
Standard		5.18	5.83

4-Methylaniline (*p*-toluidine) (C_7H_9N Mol.Wt.: 107.2 Handbook: t288)

	p.ol mass author	p.ol mass weighted	p.ol vol weighted
Backman 1917	5.94	5.94	6.59
Huijer 1918	4.85	5.91	6.56
Stuiver 1958	6.92	5.63	6.28
Standard		5.83	6.48

methyl anthranylate see: **Methyl 2-aminobenzoate**

4-Methylbenzenethiol (*p*-thiocresol) (C_7H_8S Mol.Wt.: 124.2 Handbook: t546)

	p.ol mass author	p.ol mass weighted	p.ol vol weighted
Katz and Talbert	7.86	8.06	8.77
Standard		8.06	8.77

Methyl benzoate ($C_8H_8O_2$ Mol.Wt.: 136.2 Handbook: b1286)

	p.ol mass author	p.ol mass weighted	p.ol vol weighted
Appell 1969	8.61	6.22	6.97
Standard		6.22	6.97

3-Methylbutanal (isovaleraldehyde) ($C_5H_{10}O$ Mol.Wt.: 86.1 Handbook: b2481)

	p.ol mass author	p.ol mass weighted	p.ol vol weighted
Bedborough and Trott 1979	8.79	8.09	8.65
Standard		8.09	8.65

3-Methylbutane-1-thiol (isoamyl mercaptan) ($C_5H_{12}S$ Mol.Wt.: 104.2 Handbook: b2648)

	p.ol mass author	p.ol mass weighted	p.ol vol weighted
Katz and Talbert 1930	8.74	8.94	9.58
Standard		8.94	9.58

2-Methylbutane-2-thiol (tert.amyl mercaptan) ($C_5H_{12}S$ Mol.Wt.: 104.2 Handbook: b2647*)

	p.ol mass author	p.ol mass weighted	p.ol vol weighted
Deadman and Prigg 1959	8.82	8.50	9.14
Standard		8.50	9.14

Methyl butanoate ($C_5H_{10}O_2$ Mol.Wt.: 102.1 Handbook: b2674)

	p.ol mass author	p.ol mass weighted	p.ol vol weighted
Tempelaar 1913	7.11	7.70	8.33
Standard		7.70	8.33

2-Methylbutanoic acid (α-methylbutyric acid) ($C_5H_{10}O_2$ Mol.Wt.: 102.1 Handbook: b2797)

	p.ol mass author	p.ol mass weighted	p.ol vol weighted
Punter 1983	7.70	8.10	8.73
Standard		8.10	8.73

3-Methylbutanoic acid (isovaleric acid) ($C_5H_{10}O_2$ Mol.Wt.: 102.1 Handbook: b2802)

	p.ol mass author	p.ol mass weighted	p.ol vol weighted
Backman 1917	7.02	7.02	7.65
Hesse 1926	5.52	7.09	7.72
Schwarz 1955	8.10	8.16	8.79
Kerka and Humphreys 1956	8.30	7.92	8.55
Amoore 1977	8.38	7.62	8.25
Hendriks 1979	9.08	8.75	9.38
Anon 1980	9.66	8.69	9.32
Punter 1983	8.22	8.62	9.25
Standard		7.98	8.61

2-Methylbutan-1-ol ($C_5H_{12}O$ Mol.Wt.: 88.2 Handbook: b2893)

	p.ol mass author	p.ol mass weighted	p.ol vol weighted
Backman 1917	5.81	5.81	6.38
Hellman and Small 1973, 1974	6.85	6.33	6.90
Standard		6.07	6.64

(l)2-Methylbutan-1-ol ($C_5H_{12}O$ Mol.Wt.: 88.2 Handbook: b2894)

	p.ol mass author	p.ol mass weighted	p.ol vol weighted
Passy 1892*c*	6.22	6.37	6.94
Standard		6.37	6.94

3-Methylbutan-1-ol (isoamyl alcohol) ($C_5H_{12}O$ Mol.Wt.: 88.2 Handbook: b2895)

	p.ol mass author	p.ol mass weighted	p.ol vol weighted
Passy 1892*c*	7.00	7.15	7.72
Backman 1917	6.58	6.58	7.15
Jung 1936	7.09	6.62	7.19
Standard		6.79	7.35

2-Methylbutan-2-ol (tert.amyl alcohol) ($C_5H_{12}O$ Mol.Wt.: 88.2 Handbook: b2896)

	p.ol mass author	p.ol mass weighted	p.ol vol weighted
Passy 1892*c*	4.54	4.69	5.26
Backman 1917	5.61	5.61	6.18
Standard		5.15	5.72

3-Methylbutan-2-one (methyl isopropyl ketone) ($C_5H_{10}O$ Mol.Wt.: 86.1 Handbook: b2926)

	p.ol mass author	p.ol mass weighted	p.ol vol weighted
Backman 1917	4.79	4.79	5.35
Standard		4.79	5.35

methyl tert.butyl ketone see: **3,3-Dimethylbutan-2-one**

3-Methylbutylthio-3-methylbutane (diisoamyl sulfide) ($C_{10}H_{22}S$ Mol.Wt.: 174.4 Handbook: s298)

	p.ol mass author	p.ol mass weighted	p.ol vol weighted
Katz and Talbert 1930	7.68	7.88	8.74
Standard		7.88	8.74

α-methylbutyric acid see: **2-Methylbutanoic acid**
methyl cellosolve see: **2-Methoxyethanol**
methyl cellosolve acetate see: **2-Methoxyethyl acetate**
methylchloroform see: **1,1,1-Trichlorethane**
methyl p-cresol see: **1-Methoxy-4-methylbenzene**

Methyldithioethane (methylethyl disulfide) ($C_3H_8S_2$ Mol.Wt.: 108.2 Handbook: d264*)

	p.ol mass author	p.ol mass weighted	p.ol vol weighted
Wilby 1969	7.21	7.13	7.78
Standard		7.13	7.78

Methyldithiomethane (dimethyl disulfide) ($C_2H_6S_2$ Mol.Wt.: 94.2 Handbook: d242)

	p.ol mass author	p.ol mass weighted	p.ol vol weighted
Wilby 1969	7.54	7.46	8.05
Bedborough and Trott 1979	7.34	6.64	7.23
Anon 1980	8.83	7.86	8.45
Standard		7.32	7.91

Methyl dodecanoate ($C_{13}H_{26}O_2$ Mol.Wt.: 214.4 Handbook: d303)

	p.ol mass author	p.ol mass weighted	p.ol vol weighted
Backman 1917	8.71	8.71	9.66
Standard		8.71	9.66

methylene chloride see: **Dichloromethane**

3,4-Methylenedioxybenzaldehyde (piperonal,heliotropine) ($C_8H_6O_3$ Mol.Wt.: 150.1 Handbook: b160)

	p.ol mass author	p.ol mass weighted	p.ol vol weighted
Passy 1892*a*, 1892*b*	7.15	7.30	8.10
Backman 1917	7.73	7.73	8.53
Henning 1924	8.00	7.19	7.99
Van Anrooij 1931	9.00	8.26	9.06
Baldus 1936	7.61	6.51	7.31
Appel 1969	10.52	8.13	8.93
Standard		7.52	8.32

Methyl ethoxydithioformate (methyl xanthate) ($C_4H_8OS_2$ Mol.Wt.: 136.2 Handbook: x11*)

	p.ol mass author	p.ol mass weighted	p.ol vol weighted
Deadman and Prigg 1959	8.10	7.78	8.53
Standard		7.78	8.53

methylethyl disulfide see: **Methyldithioethane**
methyl ethyl ketone see: **2-Butanone**

Methyl formate ($C_2H_4O_2$ Mol.Wt.: 60.1 Handbook: f141)

	p.ol mass author	p.ol mass weighted	p.ol vol weighted
Backman 1917	3.76	3.76	4.16
May 1966	2.30	3.49	3.89
Standard		3.63	4.03

methyl glycol see: **2-Methoxyethanol**

Methyl heptanoate ($C_8H_{16}O_2$ Mol.Wt.: 144.2 Handbook: h152)

	p.ol mass author	p.ol mass weighted	p.ol vol weighted
Backman 1917	6.38	6.38	7.16
Standard		6.38	7.16

methylheptenone see: **6-Methylhept-5-en-2-one**

6-Methylhept-5-en-2-one (methylheptenone) ($C_8H_{14}O$ Mol.Wt.: 126.2 Handbook: h212)

	p.ol mass author	p.ol mass weighted	p.ol vol weighted
Tempelaar 1913	6.52	7.11	7.83
Backman 1917	6.28	6.28	7.00
Standard		6.69	7.42

methyl heptyne carbonate see: **Methyl 2-octynoate**

5-Methylhexan-2-one (methyl isoamyl ketone) ($C_7H_{14}O$ Mol.Wt.: 114.2 Handbook: h531)

	p.ol mass author	p.ol mass weighted	p.ol vol weighted
Hellman and Small 1974	7.22	6.70	7.38
Standard		6.70	7.38

Methyl 2-hydroxypropanoate (methyl lactate) ($C_4H_8O_3$ Mol.Wt.: 104.1 Handbook: p1472

	p.ol mass author	p.ol mass weighted	p.ol vol weighted
Backman 1917	4.68	4.68	5.32
Standard		4.68	5.32

3-Methylindole (skatole) (C_9H_9N Mol.Wt.: 131.2 Handbook: i72)

	p.ol mass author	p.ol mass weighted	p.ol vol weighted
Hermanides 1909	9.45	8.90	9.64
Zwaardemaker 1914	9.40	9.32	10.06
Katz and Talbert 1930	7.00	7.20	7.94
Van Anrooij 1931	9.11	8.37	9.11
Logtanberg 1978	9.30	8.77	9.51
Standard		8.51	9.25

methyl isoamyl ketone see: **5-Methylhexan-2-one**
methyl isobutyl ketone see: **4-Methylpentan-2-one**
methyl isopropyl ketone see: **3-Methylbutan-2-one**

Methyl isothiocyanate (C_2H_3NS Mol.Wt.: 73.1 Handbook: i231)

	p.ol mass author	p.ol mass weighted	p.ol vol weighted
Allison and Katz 1919	4.83	7.41	7.89
Standard		7.41	7.89

methyl isovalerate see: **Methyl 3-methylbutanoate**
methyl lactate see: **Methyl 2-hydroxypropanoate**
methyl methacrylate see: **Methyl 2-methylprop-2-enoate**

Methyl 3-methylbutanoate (methyl isovalerate) ($C_6H_{12}O_2$ Mol.Wt.: 116.2 Handbook: b2809)

	p.ol mass author	p.ol mass weighted	p.ol vol weighted
Backman 1917	6.80	6.80	7.49
Standard		6.80	7.49

Methyl 2-methylprop-2-enoate (methyl methacrylate) ($C_5H_8O_2$ Mol.Wt.: 100.1 Handbook: p1799)

	p.ol mass author	p.ol mass weighted	p.ol vol weighted
Leonardos *et al.* 1969	6.07	6.06	6.68
Hellman and Small 1973, 1974	6.70	6.18	6.80
Anon 1980	6.21	5.24	5.86
Standard		5.83	6.45

1-Methyl-3-nitrobenzene ($C_7H_7NO_2$ Mol.Wt.: 137.1 Handbook: t555)

	p.ol mass author	p.ol mass weighted	p.ol vol weighted
Punter 1983	6.60	7.00	7.76
Standard		7.00	7.76

4-Methyl-2-nitrophenol ($C_7H_7NO_3$ Mol.Wt.: 153.1 Handbook: p679*)

	p.ol mass author	p.ol mass weighted	p.ol vol weighted
Backman 1917	6.87	6.87	7.68
Standard		6.87	7.68

Methyl nonanoate ($C_{10}H_{20}O_2$ Mol.Wt.: 172.3 Handbook: n538)

	p.ol mass author	p.ol mass weighted	p.ol vol weighted
Backman 1917	7.35	7.35	8.21
Standard		7.35	8.21

Methyl 2-octynoate (methyl heptyne carbonate) ($C_9H_{16}O_2$ Mol.Wt.: 156.2 Handbook: o228*)

	p.ol mass author	p.ol mass weighted	p.ol vol weighted
Appell 1969	10.70	8.31	9.12
Standard		8.31	9.12

2-Methylpentanal ($C_6H_{12}O$ Mol.Wt.: 100.2 Handbook: p80)

	p.ol mass author	p.ol mass weighted	p.ol vol weighted
Hellman and Small 1974	6.40	5.88	6.50
Standard		5.88	6.50

4-Methylpentanoic acid (isocaproic acid) ($C_6H_{12}O_2$ Mol.Wt.: 116.2 Handbook: p291)

	p.ol mass author	p.ol mass weighted	p.ol vol weighted
Backman 1917	6.41	6.41	7.10
Punter 1983	7.43	7.83	8.52
Standard		7.12	7.81

2-Methylpentan-1-ol ($C_6H_{14}O$ Mol.Wt.: 102.2 Handbook: p330)

	p.ol mass author	p.ol mass weighted	p.ol vol weighted
Hellman and Small 1974	7.00	6.48	7.11
Standard		6.48	7.11

4-Methylpentan-2-ol ($C_6H_{14}O$ Mol.Wt.: 102.2 Handbook: p336)

	p.ol mass author	p.ol mass weighted	p.ol vol weighted
Hellman and Small 1974	5.85	5.33	5.96
Standard		5.33	5.96

4-Methylpentan-2-one (methyl isobutyl ketone) ($C_6H_{12}O$ Mol.Wt.: 100.2 Handbook: p365)

	p.ol mass author	p.ol mass weighted	p.ol vol weighted
Backman 1917	6.16	6.16	6.78
May 1966	4.50	5.69	6.31
Stone *et al.* 1967, 1972	5.51	5.24	5.86
Leonardos *et al.* 1969	5.72	5.71	6.33
Hellman and Small 1974	6.40	5.88	6.50
Anon 1980	6.16	5.19	5.81
Standard		5.64	6.27

4-Methylpent-3-en-2-one (mesityl oxide) ($C_6H_{10}O$ Mol.Wt.: 98.2 Handbook: p441)

	p.ol mass author	p.ol mass weighted	p.ol vol weighted
Hellman and Small 1974	7.16	6.64	7.25
Standard		6.64	7.25

4-Methylpent-2-yl acetate ($C_8H_{16}O_2$ Mol.Wt.: 144.2 Handbook: a237)

	p.ol mass author	p.ol mass weighted	p.ol vol weighted
Hellman and Small 1974	6.40	5.88	6.66
Standard		5.88	6.66

2-Methylphenol (*o*-cresol (C_7H_8O Mol.Wt.: 108.2 Handbook: t498)

	p.ol mass author	p.ol mass weighted	p.ol vol weighted
Backman 1917	8.40	8.40	9.05
Stuiver 1958	9.40	8.11	8.76
Kendall *et al.* 1968	8.56	8.15	8.80
Anon 1980	8.77	7.80	8.45
Standard		8.11	8.77

3-Methylphenol (*m*-cresol) (C_7H_8O Mol.Wt.: 108.2 Handbook: t499)

	p.ol mass author	p.ol mass weighted	p.ol vol weighted
Backman 1917	9.10	9.10	9.75
Stuiver 1958	9.40	8.11	8.76
Nader 1958	8.56	8.31	8.96
Anon 1980	9.25	8.28	8.93
Standard		8.45	9.10

4-Methylphenol (*p*-cresol) (C_7H_8O Mol.Wt.: 108.2 Handbook: t500)

	p.ol mass author	p.ol mass weighted	p.ol vol weighted
Backman 1917	7.46	7.46	8.11
Baldus 1936	7.91	6.81	7.46
Stuiver 1958	10.31	9.02	9.67
Leonardos *et al.* 1969	8.36	8.35	9.00
Anon 1980	9.75	8.78	9.43
Standard		8.08	8.73

4-Methylphenyl acetate (4-tolvl acetate, p-cresyl acetate) ($C_9H_{10}O_2$ Mol.Wt.: 150.2 Handbook: a600)

	p.ol mass author	p.ol mass weighted	p.ol vol weighted
Appell 1969	10.70	8.31	9.11
Standard		8.31	9.11

2-Methylpropanal (isobutyraldhyde) (C_4H_8O Mol.Wt.: 72.1 Handbook: p1098*)

	p.ol mass author	p.ol mass weighted	p.ol vol weighted
Hellman and Small 1973, 1974	6.85	6.33	6.81
Amoore 1977	7.82	7.06	7.54
Hendriks 1979	7.66	7.33	7.81
Standard		6.91	7.39

2-Methylpropane (isobutane) (C_4H_{10} Mol.Wt.: 58.1 Handbook: p1199*)

	p.ol mass author	p.ol mass weighted	p.ol vol weighted
Mulins 1955	2.87	4.62	5.00
Standard		4.62	5.00

2-Methylpropane-1-thiol (isobutyl mercaptan) ($C_4H_{10}S$ Mol.Wt.: 90.2 Handbook: p1295)

	p.ol mass author	p.ol mass weighted	p.ol vol weighted
Allison and Katz 1919	5.09	7.67	8.25
Stuiver 1958	10.37	9.08	9.66
Wilby 1969	8.44	8.36	8.94
Standard		8.37	8.95

2-Methylpropane-2-thiol (tert.butyl mercaptan) ($C_4H_{10}S$ Mol.Wt.: 90.2 Handbook: p1296)

	p.ol mass author	p.ol mass weighted	p.ol vol weighted
Stuiver 1958	10.04	8.75	9.33
Nevers and Oister 1965	8.72	8.72	9.30
Kniebes *et al.* 1969	8.72	8.52	9.10
Wilby 1969	9.48	9.40	9.98
Blanchard 1976	8.15	8.92	9.50
Moschandreas and Jones 1983	9.04	9.09	9.67
Standard		8.90	9.48

Methyl propanoate ($C_4H_8O_2$ Mol.Wt.: 88.1 Handbook: p1324)

	p.ol mass author	p.ol mass weighted	p.ol vol weighted
Backman 1917	4.95	4.95	5.52
Standard		4.95	5.52

2-Methylpropanoic acid (isobutyric acid) ($C_4H_8O_2$ Mol.Wt.: 88.1 Handbook: p1518)

	p.ol mass author	p.ol mass weighted	p.ol vol weighted
Anon 1980	8.30	7.33	7.90
Punter 1983	6.55	6.95	7.52
Standard		7.14	7.71

2-Methylpropan-1-ol (isobutyl alcohol) ($C_4H_{10}O$ Mol.Wt.: 74.1 Handbook: p1633*)

	p.ol mass author	p.ol mass weighted	p.ol vol weighted
Passy 1892*c*	6.00	6.15	6.64
Zwaardemaker 1914	3.30	3.22	3.71
Backman 1917	6.55	6.55	7.04
Jones 1955*b*	4.51	6.02	6.51
May 1966	3.92	5.11	5.60
Laffort and Dravnieks 1973	5.52	5.93	6.42
Hellman and Small 1973, 1974	5.70	5.18	5.67
Anon 1980	7.44	6.47	6.96
Punter 1983	5.26	5.66	6.15
Standard		5.59	6.08

2-Methylpropan-2-ol (tert.butyl alcohol) ($C_4H_{10}O$ Mol.Wt.: 74.1 Handbook: p1633)

	p.ol mass author	p.ol mass weighted	p.ol vol weighted
Passy 1892*c*	4.85	5.00	5.49
Backman 1917	4.42	4.42	4.91
Jones 1955*b*	3.12	4.63	5.12
Dravnieks and Laffort 1972	4.15	4.25	4.74
Dravnieks 1974	2.54	3.74	4.23
Standard		4.41	4.90

2-Methylpropene (isobutylene) (C_4H_8 Mol.Wt.: 56.1 Handbook: p1747)

	p.ol mass author	p.ol mass weighted	p.ol vol weighted
Katz and Talbert 1930	5.52	5.72	6.09
Mullins 1955	2.31	4.06	4.43
Anon 1980	4.82	3.85	4.22
Standard		4.54	4.91

Methyl 2-propenoate (methyl acrylate) ($C_4H_6O_2$ Mol.Wt.: 86.1 Handbook: p1775)

	p.ol mass author	p.ol mass weighted	p.ol vol weighted
Anon 1980	8.00	7.03	7.58
Standard		7.03	7.58

methyl propyl ketone see: **2-Pentanone**

2-Methylpyridine (2-picoline) (C_6H_7N Mol.Wt.: 93.1 Handbook: p2019)

	p.ol mass author	p.ol mass weighted	p.ol vol weighted
Hellman and Small 1974	7.30	6.78	7.37
Standard		6.78	7.37

Methyl salicylate ($C_8H_8O_3$ Mol.Wt.: 152.2 Handbook: b1671)

	p.ol mass author	p.ol mass weighted	p.ol vol weighted
Tempelaar 1913	6.02	6.61	7.41
Backman 1917	6.07	6.07	6.87
Allison and Katz 1919	4.00	6.58	7.38
Baldus 1936	8.70	7.60	8.40
Jung 1936	7.64	7.17	7.97
Jones 1954	4.80	6.48	7.28
Jones 1955*b*	4.65	6.16	6.96

Methyl salicylate *continued*

Kerka and Humphreys 1956	7.22	6.84	7.64
Nader 1958	6.88	6.63	7.43
Köster 1971	3.98	5.48	6.28
Standard		6.56	7.36

α-methylstyrene see: **Isopropenylbenzene**

Methyl thiocyanate (C_2H_3NS Mol.Wt.: 73.1 Handbook: t175)

	p.ol mass author	p.ol mass weighted	p.ol vol weighted
Katz and Talbert 1930	6.13	6.33	6.81
Standard		6.33	6.81

Methylthioethane (ethyl methyl sulfide) (C_3H_8S Mol.Wt.: 76.2 Handbook: s341)

	p.ol mass author	p.ol mass weighted	p.ol vol weighted
Moschandreas and Jones 1983	7.87	7.92	8.42
Standard		7.92	8.42

Methyl 2-thiolacetate (C_3H_6OS Mol.Wt.: 106.1 Handbook: a591*)

	p.ol mass author	p.ol mass weighted	p.ol vol weighted
Deadman and Prigg 1959	8.30	7.98	8.63
Standard		7.98	8.63

Methylthiomethane (dimethyl sulfide) (C_2H_6S Mol.Wt.: 62.1 Handbook: s313)

	p.ol mass author	p.ol mass weighted	p.ol vol weighted
Katz and Talbert 1930	8.03	8.23	8.64
Nevers 1965 and Oister	8.46	8.46	8.87
Guadagni 1966	8.53	8.29	8.70
Leonardos *et al.* 1969	8.61	8.60	9.01
Wilby 1969	8.20	8.12	8.53
Laffort and Dravnieks 1973	7.86	8.27	8.68
Cormack *et al.* 1974	8.13	7.76	8.17
Nishida *et al.* 1975	7.90	8.05	8.46
Anon 1980	9.53	8.56	8.97
Moschandreas and Jones 1983	7.97	8.02	8.43
Standard		8.23	8.65

2-Methyl-3,3,3-trichloropropan-1-ol ($C_4H_7OCl_3$ Mol.Wt.: 177.5 Handbook: p1638*)

	p.ol mass author	p.ol mass weighted	p.ol vol weighted
Backman 1917	5.21	5.21	6.08
Standard		5.21	6.08

Methyltrithiomethane (dimethyl trisulfide) ($C_2H_6S_3$ Mol.Wt.: 126.3 Handbook: t746*)

	p.ol mass author	p.ol mass weighted	p.ol vol weighted
Wilby 1969	8.14	8.06	8.78
Standard		8.06	8.78

2-Methylundecane-2-thiol (tert.dodecyl mercaptan) ($C_{12}H_{26}S$ Mol.Wt.: 202.4 Handbook: h21*)

	p.ol mass author	p.ol mass weighted	p.ol vol weighted
Blinova 1965	5.22	5.79	6.72
Standard		5.79	6.72

methyl xanthate see: **Methyl ethoxydithioformate**

Morpholine (C_4H_9NO Mol.Wt.: 87.1 Handbook: m408)

	p.ol mass author	p.ol mass weighted	p.ol vol weighted
Hellman and Small 1973, 1974	7.40	6.88	7.44
Standard		6.88	7.44

musk ambrette see: **1-tert.Butyl-3,5-dinitro-2-methoxy-4-methylbenzene**
musk xylene see: **1-tert.Butyl-3,5-dimethyl-2,4,6-trinitrobenzene**

Naphthalene ($C_{10}H_8$ Mol.Wt.: 128.2 Handbook: n14)

	p.ol mass author	p.ol mass weighted	p.ol vol weighted
Backman 1917	7.28	7.28	8.01
Mitsumoto 1926	5.38	6.42	7.15
Hesse 1927	6.52	8.14	8.87
Morimura 1934	5.37	6.54	7.27
Punter 1983	6.70	7.10	7.83
Standard		7.10	7.83

1-naphthol see: **1-Hydroxynaphthalene**
2-naphthol see: **2-Hydroxynaphthalene**
1-naphthylamine see: **1-Aminonaphthalene**
2-naphthylamine see: **2-Aminonaphthalene**
nerol see: **3,7-Dimetylocta-cis-2,6-dien-1-ol**

2-Nitrobenzaldehyde ($C_7H_5NO_3$ Mol.Wt.: 151.1 Handbook: b162)

	p.ol mass author	p.ol mass weighted	p.ol vol weighted
Backman 1917	5.35	5.35	6.15
Standard		5.35	6.15

3-Nitrobenzaldehyde ($C_7H_5NO_3$ Mol.Wt.: 151.1 Handbook: b165)

	p.ol mass author	p.ol mass weighted	p.ol vol weighted
Backman 1917	5.52	5.52	6.32
Standard		5.52	6.32

Nitrobenzene ($C_6H_5NO_2$ Mol.Wt.: 123.1 Handbook: b891)

	p.ol mass author	p.ol mass weighted	p.ol vol weighted
Hermanides 1909	7.39	6.84	7.55
Zwaardemaker 1914	7.39	7.31	8.02
Backman 1917	6.31	6.31	7.02
Allison and Katz 1919	3.84	6.42	7.13
Henning 1924	8.19	7.38	8.09
Katz and Talbert 1930	5.02	5.22	5.93
Van Anrooij 1931	7.72	6.98	7.69
Janicek *et al.* 1960	4.72	6.30	7.01
Gavaudan and Poussel 1966	6.82	6.36	7.07
Leonardos *et al.* 1969	7.62	7.61	8.32
Randebrock 1971	8.70	6.85	7.56
Naus 1982	6.70	6.22	6.93
Standard		6.65	7.36

Nitrogen dioxide (NO_2 Mol.Wt.: 46.0 Handbook: I, n130)

	p.ol mass author	p.ol mass weighted	p.ol vol weighted
Beck 1959	6.35	6.08	6.36
Shalamberidze 1967	6.64	6.55	6.83
Knuth 1973	6.96	6.73	7.01
Standard		6.45	6.73

2-Nitrophenol ($C_6H_5NO_3$ Mol.Wt.: 139.1 Handbook: p679)

	p.ol mass author	p.ol mass weighted	p.ol vol weighted
Backman 1917	6.93	6.93	7.69
Stuiver 1958	8.92	7.63	8.39
Standard		7.28	8.04

3-Nitrophenol ($C_6H_5NO_3$ Mol.Wt.: 139.1 Handbook: p680)

	p.ol mass author	p.ol mass weighted	p.ol vol weighted
Stuiver 1958	5.53	4.24	5.00
Standard		4.24	5.00

4-Nitrophenol ($C_6H_5NO_3$ Mol.Wt.: 139.1 Handbook: p681)

	p.ol mass author	p.ol mass weighted	p.ol vol weighted
Stuiver 1958	5.64	4.35	5.11
Standard		4.35	5.11

1-Nitropropane ($C_3H_7NO_2$ Mol.Wt.: 89.1 Handbook: p1210)

	p.ol mass author	p.ol mass weighted	p.ol vol weighted
Dravnieks and Laffort 1972	4.55	4.65	5.22
Dravnieks 1974	3.29	4.49	5.06
Standard		4.57	5.14

Nitrosobenzene (C_6H_5NO Mol.Wt.: 107.1 Handbook: b894)

	p.ol mass author	p.ol mass weighted	p.ol vol weighted
Randebrock 1971	6.82	4.97	5.62
Standard		4.97	5.62

trans-2,trans-4-Nonadienal ($C_9H_{14}O$ Mol.Wt.: 138.2 Handbook: n515*)

	p.ol mass author	p.ol mass weighted	p.ol vol weighted
Teranishi *et al.* 1974	9.60	9.51	10.27
Hall and Andersson 1983	9.40	9.87	10.63
Standard		9.69	10.45

trans-2,cis-6-Nonadienal ($C_9H_{14}O$ Mol.Wt.: 138.2 Handbook: n515*)

	p.ol mass author	p.ol mass weighted	p.ol vol weighted
Teranishi *et al.* 1974	10.22	10.13	10.89
Standard		10.13	10.89

Nonanal ($C_9H_{18}O$ Mol.Wt.: 142.2 Handbook: n517)

	p.ol mass author	p.ol mass weighted	p.ol vol weighted
Pliska and Janicek 1965	5.77	8.57	9.34
Randebrock 1971	9.53	7.68	8.45
Teranishi *et al.* 1974	7.53	7.44	8.21
Hall and Andersson 1983	7.35	7.82	8.59
Standard		7.87	8.65

Nonane (C_9H_{20} Mol.Wt.: 128.3 Handbook: n519)

	p.ol mass author	p.ol mass weighted	p.ol vol weighted
Mullins 1955	3.96	5.71	6.44
Laffort and Dravnieks 1973	4.22	4.63	5.36
Standard		5.17	5.90

Nonanoic acid (pelargonic acid) ($C_9H_{18}O_2$ Mol.Wt.: 158.2 Handbook: n534)

	p.ol mass author	p.ol mass weighted	p.ol vol weighted
Passy 1893*b*, 1893*c*	7.70	7.73	8.55
Backman 1917	8.64	8.64	9.46
Punter 1983	6.92	7.32	8.14
Standard		7.90	8.72

1-Nonanol ($C_9H_{20}O$ Mol.Wt.: 144.3 Handbook: n543)

	p.ol mass author	p.ol mass weighted	p.ol vol weighted
Pliska 1962	4.39	6.37	7.15
Standard		6.37	7.15

2-Nonanone ($C_9H_{18}O$ Mol.Wt.: 142.2 Handbook: n552)

	p.ol mass author	p.ol mass weighted	p.ol vol weighted
Teranishi *et al.* 1974	7.13	7.04	7.81
Hall and Andersson 1983	5.77	6.24	7.01
Standard		6.64	7.41

trans-2-Nonenal ($C_9H_{16}O$ Mol.Wt.: 140.2 Handbook: n556*)

	p.ol mass author	p.ol mass weighted	p.ol vol weighted
Teranishi *et al.* 1974	9.30	9.21	9.98
Hall and Andersson 1983	8.44	8.91	9.68
Standard		9.06	9.83

trans-2-Nonen-4-one ($C_9H_{16}O$ Mol.Wt.: 140.2 Handbook: n557*)

	p.ol mass author	p.ol mass weighted	p.ol vol weighted
Teranishi *et al.* 1974	8.57	8.48	9.25
Standard		8.48	9.25

trans-2,trans-4-Octadienal ($C_8H_{12}O$ Mol.Wt.: 124.2 Handbook: o98*)

	p.ol mass author	p.ol mass weighted	p.ol vol weighted
Hall and Andersson 1983	7.92	8.39	9.10
Standard		8.39	9.10

Octanal ($C_8H_{16}O$ Mol.Wt.: 128.2 Handbook: o110)

	p.ol mass author	p.ol mass weighted	p.ol vol weighted
Pliska and Janicek 1968	5.30	8.10	8.83
Teranishi *et al.* 1974	7.82	7.73	8.46
Hall and Andersson 1983	8.11	8.58	9.31
Standard		8.14	8.87

Octane (C_8H_{18} Mol.Wt.: 114.2 Handbook: o112)

	p.ol mass author	p.ol mass weighted	p.ol vol weighted
Jones 1955*b*	3.26	4.77	5.45
May 1966	3.15	4.34	5.02
Laffort and Dravnieks 1973	4.15	4.56	5.24
Standard		4.56	5.24

Octanoic acid (caprylic acid) ($C_8H_{16}O_2$ Mol.Wt.: 144.2 Handbook: o161)

	p.ol mass author	p.ol mass weighted	p.ol vol weighted
Passy 1893*b*, 1893*c*	7.30	7.33	8.11
Backman 1917	9.52	9.52	10.30
Jones 1955*b*	5.49	7.00	7.78
Punter 1983	6.22	6.62	7.40
Standard		7.62	8.40

1-Octanol ($C_8H_{18}O$ Mol.Wt.: 130.2 Handbook: o186)

	p.ol mass author	p.ol mass weighted	p.ol vol weighted
Backman 1917	8.20	8.20	8.93
Rocen 1920	8.31	8.86	9.59
Mullins 1955	5.27	7.02	7.75
Pliska 1962	5.05	7.03	7.76
Stone *et al.* 1967	7.31	7.04	7.77
Cain 1969	6.31	7.86	8.59
Punter 1983	6.14	6.54	7.27
Standard		7.50	8.24

2-Octanol ($C_8H_{18}O$ Mol.Wt.: 130.2 Handbook: o188)

	p.ol mass author	p.ol mass weighted	p.ol vol weighted
Passy 1892*c*	7.31	7.46	8.19
Backman 1917	7.85	7.85	8.58
Punter 1983	6.86	7.26	7.99
Standard		7.52	8.25

(d)2-Octanol ($C_8H_{18}O$ Mol.Wt.: 130.2 Handbook: o187)

	p.ol mass author	p.ol mass weighted	p.ol vol weighted
Stuiver 1958	8.96	7.67	8.40
Punter 1983	7.18	7.58	8.31
Standard		7.62	8.36

(1)2-Octanol ($C_8H_{18}O$ Mol.Wt.: 130.2 Handbook: o189)

	p.ol mass author	p.ol mass weighted	p.ol vol weighted
Stuiver 1958	9.53	8.24	8.97
Punter 1983	6.66	7.06	7.79
Standard		7.65	8.38

2-Octanone ($C_8H_{16}O$ Mol.Wt.: 128.2 Handbook: o197)

	p.ol mass author	p.ol mass weighted	p.ol vol weighted
Backman 1917	7.11	7.11	7.84
Stone 1963*b*, 1963*c*	6.11	6.89	7.62
Hall and Andersson 1983	6.64	7.11	7.84
Standard		7.04	7.77

3-Octanone ($C_8H_{16}O$ Mol.Wt.: 128.2 Handbook: o198)

	p.ol mass author	p.ol mass weighted	p.ol vol weighted
Appell 1969	8.88	6.49	7.22
Standard		6.49	7.22

trans-2-Octenal ($C_8H_{14}O$ Mol.Wt.: 126.2 Handbook: o208*)

	p.ol mass author	p.ol mass weighted	p.ol vol weighted
Teranishi *et al.* 1974	8.05	7.96	8.68
Eriksson *et al.* 1976	7.85	8.14	8.86
Hall and Andersson 1983	7.33	7.80	8.52
Standard		7.97	8.69

1-Octene (C_8H_{16} Mol.Wt.: 112.2 Handbook: o208)

	p.ol mass author	p.ol mass weighted	p.ol vol weighted
Dravnieks and Laffort 1972	6.48	6.58	7.25
Dravnieks 1974	5.30	6.50	7.17
Standard		6.54	7.21

2-Octene (C_8H_{16} Mol.Wt.: 112.2 Handbook: o209)

	p.ol mass author	p.ol mass weighted	p.ol vol weighted
Dravnieks and Laffort 1972	6.30	6.40	7.07
Dravnieks 1974	5.30	6.50	7.17
Standard		6.45	7.12

1-Octen-3-ol ($C_{18}H_{16}O$ Mol.Wt.: 128.2 Handbook: o215*)

	p.ol mass author	p.ol mass weighted	p.ol vol weighted
Hall and Andersson 1983	7.32	7.79	8.52
Standard		7.79	8.52

2-Octyl acetate (sec. octyl acetate) ($C_{10}H_{20}O_2$ Mol.Wt.: 172.3 Handbook: a249)

	p.ol mass author	p.ol mass weighted	p.ol vol weighted
Backman 1917	7.54	7.54	8.40
Standard		7.54	8.40

sec.octyl acetate see: **2-Octyl acetate**

2-Octyne (C_8H_{14} Mol.Wt.: 110.2 Handbook: o222)

	p.ol mass author	p.ol mass weighted	p.ol vol weighted
Dravnieks and Laffort 1972	7.53	7.63	8.29
Dravnieks 1974	5.70	6.90	7.56
Standard		7.26	7.93

oenanthic acid see: **Heptanoic acid**

2-Oxapropyl salicylate (methoxymethyl salicylate,mesotan) ($C_9H_{10}O_4$ Mol.Wt.: 182.2 Handbook: b1669)

	p.ol mass author	p.ol mass weighted	p.ol vol weighted
Backman 1917	5.00	5.00	5.88
Standard		5.00	5.88

Ozone (O_3 Mol.Wt.: 48.0 Handbook: I, o21)

	p.ol mass author	p.ol mass weighted	p.ol vol weighted
Beck 1959	7.40	7.13	7.43
Buchberg *et al.* 1961	6.73	6.85	7.15
Standard		6.99	7.29

Paraldehyde ($C_6H_{12}O_3$ Mol.Wt.: 132.2 Handbook: p15)

	p.ol mass author	p.ol mass weighted	p.ol vol weighted
Backman 1917	7.65	7.65	8.39
Standard		7.65	8.39

pelargonic acid see: **Nonaoic acid**

Pentanal (valeraldehyde) ($C_5H_{10}O$ Mol.Wt.: 86.1 Handbook: p77)

	p.ol mass author	p.ol mass weighted	p.ol vol weighted
Backman 1917	8.02	8.02	8.58
Teranishi *et al.* 1974	7.14	7.05	7.61
Anon 1980	8.60	7.63	8.19
Hall and Andersson 1983	7.46	7.93	8.49
Standard		7.66	8.22

Pentane (C_5H_{12} Mol.Wt.: 72.2 Handbook: p84)

	p.ol mass author	p.ol mass weighted	p.ol vol weighted
Patty and Yant 1929	2.84	3.94	4.42
Mullins 1955	2.51	4.26	4.74
Laffort and Dravnieks 1973	3.45	3.86	4.34
Standard		4.02	4.50

2,3-Pentanedione ($C_5H_8O_2$ Mol.Wt.: 100.1 Handbook: p198)

	p.ol mass author	p.ol mass weighted	p.ol vol weighted
Hall and Andersson 1983	7.20	7.67	8.29
Standard		7.67	8.29

2,4-Pentanedione ($C_5H_8O_2$ Mol.Wt.: 100.1 Handbook: p202)

	p.ol mass author	p.ol mass weighted	p.ol vol weighted
Hellman and Small 1974	7.40	6.88	7.50
Standard		6.88	7.50

1-Pentanethiol (amyl mercaptan) ($C_5H_{12}S$ Mol.Wt.: 104.2 Handbook: p209)

	p.ol mass author	p.ol mass weighted	p.ol vol weighted
Sales 1958	9.30	9.36	10.00
Kniebes *et al.* 1969	9.40	9.20	9.84
Standard		9.28	9.92

Pentanoic acid (valeric acid) ($C_5H_{10}O_2$ Mol.Wt.: 102.1 Handbook: p214)

	p.ol mass author	p.ol mass weighted	p.ol vol weighted
Passy 1893*b*, 1893*c*	8.00	8.30	8.66
Hermanides 1909	8.68	8.13	8.76
Zwaardemaker 1914	8.68	8.60	9.23
Backman 1917	7.70	7.70	8.33
Allison and Katz 1919	4.54	7.12	7.75
Henning 1924	9.10	8.29	8.92
Hesse 1926	5.04	6.61	7.24
Hesse 1927	5.52	7.14	7.77
Bocca and Battiston 1964	7.38	7.68	8.31
Punter 1983	7.22	7.62	8.25
Standard		7.69	8.32

1-Pentanol (amyl alcohol) ($C_5H_{12}O$ Mol.Wt.: 88.2 Handbook: p313)

	p.ol mass author	p.ol mass weighted	p.ol vol weighted
Backman 1917	5.96	5.96	6.53
Allison and Katz 1919	3.64	6.22	6.79
Jung 1936	6.24	5.77	6.34
Janicek *et al.* 1960	4.95	6.53	7.10
Naus 1962	5.39	4.77	5.34
Pliska and Janicek 1965	2.95	5.75	6.32
Gavaudan and Poussel 1966	6.39	5.93	6.50
May 1966	4.45	5.64	6.21
Stone *et al.* 1972	5.92	6.00	6.57
Hellman and Small 1974	6.09	5.57	6.14
Naus 1982	5.39	4.91	5.48
Punter 1983	5.69	6.09	6.66
Standard		5.77	6.33

2-Pentanone (methyl propyl ketone) ($C_5H_{10}O$ Mol.Wt.: 86.1 Handbook: p346)

	p.ol mass author	p.ol mass weighted	p.ol vol weighted
Backman 1917	4.89	4.89	5.45
May 1966	4.56	5.75	6.31
Hall and Andersson 1983	4.65	5.12	5.68
Standard		5.26	5.81

3-Pentanone (diethyl ketone) ($C_5H_{10}O$ Mol.Wt.: 86.1 Handbook: p348)

	p.ol mass author	p.ol mass weighted	p.ol vol weighted
Backman 1917	5.42	5.42	5.98
May 1966	4.48	5.67	6.23
Dravnieks 1974	5.52	6.72	7.28
Standard		5.94	6.50

1-Penten-3-ol ($C_5H_{10}O$ Mol.Wt.: 86.1 Handbook: p432)

	p.ol mass author	p.ol mass weighted	p.ol vol weighted
Hall and Andersson 1983	5.36	5.83	6.39
Standard		5.83	6.39

Pentyl acetate (amyl acetate) ($C_7H_{14}O_2$ Mol.Wt.: 130.2 Handbook: a253)

	p.ol mass author	p.ol mass weighted	p.ol vol weighted
Grijns 1919	6.05	7.29	8.02
Allison and Katz 1919	4.41	6.99	7.72
Jones 1955*b*	5.80	7.31	8.04
Gofmekler 1960	6.23	5.68	6.41
Guadagni 1966	7.31	7.07	7.80
Hendriks 1979	6.57	6.24	6.97
Laing 1982	6.03	6.68	7.41
Punter 1983	6.57	6.97	7.70
Standard		6.77	7.51

2-Pentylfuran ($C_9H_{14}O$ Mol.Wt.: 138.2 Handbook: f217*)

	p.ol mass author	p.ol mass weighted	p.ol vol weighted
Hall and Andersson 1983	6.57	7.04	7.80
Standard		7.04	7.80

Pentyl 3-Methylbutanoate (amyl isovalerate) ($C_{10}H_{20}O_2$ Mol.Wt.: 172.3 Handbook: b2810*)

	p.ol mass author	p.ol mass weighted	p.ol vol weighted
Allison and Katz 1919	4.92	7.50	8.36
Standard		7.50	8.36

Pentyl salicylate (amyl salicylate) ($C_{12}H_{16}O_3$ Mol.Wt.: 208.2 Handbook: b1678)

	p.ol mass author	p.ol mass weighted	p.ol vol weighted
Backman 1917	8.00	8.00	8.94
Baldus 1936	8.12	7.02	7.96
Standard		7.51	8.45

Pentylthiopentane (diamyl sulfide) ($C_{10}H_{22}S$ Mol.Wt.: 174.4 Handbook: s316)

	p.ol mass author	p.ol mass weighted	p.ol vol weighted
Allison and Katz 1919	6.00	8.58	9.44
Standard		8.58	9.44

perchloroethylene see: **Tetrachloroethene**

Phenacyl bromide (α-bromoacetophenone) (C_8H_7BrO Mol.Wt.: 199.1 Handbook: a653)

	p.ol mass author	p.ol mass weighted	p.ol vol weighted
Katz and Talbert 1930	6.92	7.12	8.04
Standard		7.12	8.04

Phenacyl chloride (α-chloroacetophenone) (C_8H_7ClO Mol.Wt.: 154.6 Handbook: a661)

	p.ol mass author	p.ol mass weighted	p.ol vol weighted
Katz and Talbert 1930	6.58	6.78	7.59
Standard		6.78	7.59

Phenanthrene ($C_{14}H_{10}$ Mol.Wt.: 178.2 Handbook: p464)

	p.ol mass author	p.ol mass weighted	p.ol vol weighted
Backman 1917	7.24	7.24	8.11
Standard		7.24	8.11

phenethyl alcohol see: **2-Phenylethanol**

Phenol (C_6H_6O Mol.Wt.: 94.1 Handbook: p555)

	p.ol mass author	p.ol mass weighted	p.ol vol weighted
Zwaardemaker 1914	5.40	5.32	5.91
Backman 1917	6.74	6.74	7.33
Henning 1924	5.92	5.11	5.70
Basmadzhieva and Argirova 1968	7.68	6.83	7.42
Leonardos *et al.* 1969	6.75	6.74	7.33
Takhirov 1974	7.66	6.85	7.44
Makeicheva 1978	7.57	6.47	7.06

Phenol *continued*

Anon 1980	7.34	6.37	6.96
Naus 1982	6.70	6.22	6.81
Punter 1983	6.64	7.04	7.63
Standard		6.37	6.96

Phenoxybenzene (diphenyl ether) ($C_{12}H_{10}O$ Mol.Wt.: 170.2 Handbook: e510)

	p.ol mass author	p.ol mass weighted	p.ol vol weighted
Katz and Talbert 1930	8.16	8.36	9.21
Leonardos *et al.* 1969	6.16	6.15	7.00
Appell 1969	9.77	7.38	8.23
Standard		7.30	8.15

Phenyl acetate ($C_8H_8O_2$ Mol.Wt.: 136.2 Handbook: a256)

	p.ol mass author	p.ol mass weighted	p.ol vol weighted
Jung 1936	6.70	6.23	6.98
Standard		6.23	6.98

2-Phenylacetic acid ($C_8H_8O_2$ Mol.Wt.: 136.2 Handbook: a574)

	p.ol mass author	p.ol mass weighted	p.ol vol weighted
Appell 1969	10.53	8.14	8.89
Standard		8.14	8.89

phenylcarbylamine see: **Phenyl isocyanide**

1-Phenyl-1,2-epoxyethane (styrene oxide) (C_8H_8O Mol.Wt.: 120.2 Handbook: b745)

	p.ol mass author	p.ol mass weighted	p.ol vol weighted
Hellman and Small 1974	6.52	6.00	6.70
Standard		6.00	6.70

2-Phenylethanal dimethylacetal ($C_{10}H_{14}O_2$ Mol.Wt.: 166.2
Handbook: a22*)

	p.ol mass author	p.ol mass weighted	p.ol vol weighted
Appell 1969	9.22	6.83	7.67
Standard		6.83	7.67

2-Phenylethanol (phenethyl alcohol) ($C_8H_{10}O$ Mol.Wt.: 122.2
Handbook: e389)

	p.ol mass author	p.ol mass weighted	p.ol vol weighted
Appell 1969	9.45	7.06	7.77
Standard		7.06	7.77

1-Phenylethyl acetate (styralyl acetate) ($C_{10}H_{12}O_2$ Mol.Wt.: 164.2
Handbook: a580)

	p.ol mass author	p.ol mass weighted	p.ol vol weighted
Appell 1969	9.69	7.30	8.14
Standard		7.30	8.14

Phenyl isocyanide (phenylcarbylamine) (C_7H_5NO Mol.Wt.: 119.1
Handbook: i149)

	p.ol mass author	p.ol mass weighted	p.ol vol weighted
Allison and Katz 1919	5.69	8.27	8.97
Katz and Talbert 1930	8.37	8.57	9.27
Standard		8.42	9.12

Phenyl isothiocyanate (C_7H_5NS Mol.Wt.: 135.2 Handbook: t236)

	p.ol mass author	p.ol mass weighted	p.ol vol weighted
Katz and Talbert	6.28	6.48	7.23
Stone *et al.* 1967	7.48	7.21	7.96
Standard		6.84	7.60

3-Phenylpropan-1-ol ($C_9H_{12}O$ Mol.Wt.: 136.2 Handbook: p1644)

	p.ol mass author	p.ol mass weighted	p.ol vol weighted
Schwarz 1955	8.83	8.89	9.64
Standard		8.89	9.64

3-Phenylpropenal (cinnamaldehyde) (C_9H_8O Mol.Wt.: 132.2 Handbook: c339)

	p.ol mass author	p.ol mass weighted	p.ol vol weighted
Tempelaar 1913	7.01	7.60	8.34
Backman 1917	7.96	7.96	8.70
Baldus 1936	7.70	6.60	7.34
Appell 1969	10.52	8.13	8.87
Randebrock 1971	10.83	8.98	9.72
Standard		7.85	8.59

3-Phenylprop-2-en-1-ol (cinnamic alcohol) ($C_9H_{10}O$ Mol.Wt.: 134.2 Handbook: p1822)

	p.ol mass author	p.ol mass weighted	p.ol vol weighted
Tempelaar 1913	6.72	7.31	8.06
Appell 1969	10.52	8.13	8.88
Randebrock 1971	10.88	9.03	9.78
Standard		8.16	8.91

Phenylthiobenzene (diphenyl sulfide) ($C_{12}H_{10}S$ Mol.Wt.: 186.3 Handbook: s317)

	p.ol mass author	p.ol mass weighted	p.ol vol weighted
Katz and Talbert 1930	8.59	8.79	9.68
Leonardos *et al.* 1969	7.44	7.43	8.32
Standard		8.11	9.00

phenylurethan see: **Ethyl n-phenylcarbamate**
phosgene see: **Carbonyl chloride**

Phosphine (PH_3 Mol.Wt.: 34.0 Handbook: I, p64)

	p.ol mass author	p.ol mass weighted	p.ol vol weighted
Valentin 1848, 1850	6.37	7.51	7.66
Leonardos *et al.* 1969	7.52	7.51	7.66
Standard		7.51	7.66

2-picoline see: **2-Methylpyridine**

α-Pinene ($C_{10}H_{16}$ Mol.Wt.: 136.2 Handbook: p923)

	p.ol mass author	p.ol mass weighted	p.ol vol weighted
Appell 1969	7.80	5.41	6.16
Standard		5.41	6.16

Piperidine (hexahydroazine) ($C_5H_{11}N$ Mol.Wt.: 85.2 Handbook: p952)

	p.ol mass author	p.ol mass weighted	p.ol vol weighted
Geier 1936	6.30	5.88	6.43
Standard		5.88	6.43

piperonal see: **3,4-Methylenedioxybenzaldehyde**

Propanal (C_3H_6O Mol.Wt.: 58.1 Handbook: p1083)

	p.ol mass author	p.ol mass weighted	p.ol vol weighted
Backman 1917	7.70	7.70	8.08
Pliska and Janicek 1965	3.62	6.42	6.80
Knuth 1973	7.59	7.36	7.74
Hellman and Small 1974	7.70	7.18	7.56
Teranishi *et al.* 1974	7.70	7.61	7.99
Bedborough and Trott 1979	7.86	7.16	7.54
Anon 1980	8.45	7.48	7.86
Hall and Andersson 1983	6.17	6.64	7.02
Standard		7.19	7.57

Propane (C_3H_8 Mol.Wt.: 44.1 Handbook: p1107)

	p.ol mass author	p.ol mass weighted	p.ol vol weighted
Patty and Yant 1929	1.45	2.55	2.81
Laffort and Dravnieks 1973	1.66	2.07	2.33
Standard		2.31	2.57

1,2-Propanediamine ($C_3H_{10}N_2$ Mol.Wt.: 74.1 Handbook: p1151)

	p.ol mass author	p.ol mass weighted	p.ol vol weighted
Hellman and Small 1974	7.40	6.88	7.37
Standard		6.88	7.37

1-Propanethiol (C_3H_8S Mol.Wt.: 76.2 Handbook: p1293)

	p.ol mass author	p.ol mass weighted	p.ol vol weighted
Allison and Katz 1919	5.22	7.80	8.30
Katz and Talbert 1930	8.30	8.50	9.00
Blinova 1965	7.84	8.41	8.91
Kniebes *et al.* 1969	8.80	8.60	9.10
Wilby 1969	8.64	8.56	9.06
Blanchard 1976	7.75	8.52	9.02
Standard		8.40	8.90

2-Propanethiol (isopropyl mercaptan) (C_3H_8S Mol.Wt.: 76.2 Handbook: p1294)

	p.ol mass author	p.ol mass weighted	p.ol vol weighted
Stuiver 1958	11.15	9.86	10.36
Wilby 1969	8.86	8.78	9.28
Blanchard 1976	7.75	8.52	9.02
Moschandreas and Jones 1983	8.59	8.64	9.14
Standard		8.95	9.45

Propanoic acid ($C_3H_6O_2$ Mol.Wt.: 74.1 Handbook: p1302)

	p.ol mass author	p.ol mass weighted	p.ol vol weighted
Passy 1893*b*, 1893*c*	7.30	7.33	7.82
Backman 1917	6.30	6.30	6.79
Grijns 1919	6.22	7.46	7.95
Hesse 1926	5.34	6.91	7.40
Mitsumoto 1926	5.68	6.72	7.21
Morimura 1934	5.69	6.86	7.35
Stone 1963*a*, 1963*c*	6.29	7.04	7.53
Stone and Bosley 1965	6.05	7.00	7.49
Goldenberg 1967	8.52	7.29	7.78
Hellman and Small 1974	7.10	6.58	7.07
Anon 1980	8.29	7.32	7.81
Punter 1983	6.30	6.70	7.19
Standard		6.96	7.45

1-Propanol (C_3H_8O Mol.Wt.: 60.1 Handbook: p1587)

	p.ol mass author	p.ol mass weighted	p.ol vol weighted
Passy 1892*c*	5.15	5.30	5.70
Backman 1917	5.41	5.41	5.81
Jung 1936	5.60	5.13	5.53
Jones 1955*b*	3.85	5.36	5.76
Janicek *et al.* 1960	3.27	4.85	5.25
Pliska and Janicek 1965	1.60	4.40	4.80
Guadagni 1966	5.04	4.80	5.20
May 1966	4.10	5.29	5.69
Cain 1969	3.18	4.73	5.13
Corbit and Engen 1971	4.31	5.27	5.67
Dravnieks and Laffort 1972	4.49	4.59	4.99
Stone *et al.* 1972	5.55	5.63	6.03
Dravnieks 1974	4.00	5.20	5.60
Hellman and Small 1974	7.12	6.60	7.00
Laing 1975	4.00	5.38	5.78
Naus 1982	5.70	5.22	5.62
Punter 1983	5.23	5.63	6.03
Standard		5.22	5.62

2-Propanol (isopropyl alcohol) (C_3H_8O Mol.Wt.: 60.1 Handbook: p1588)

	p.ol mass author	p.ol mass weighted	p.ol vol weighted
Passy 1892*c*	4.40	4.55	4.95
Backman 1917	4.68	4.68	5.08
Jung 1936	4.95	4.48	4.88
Scherberger *et al.* 1958	3.30	4.72	5.12
May 1966	4.04	5.23	5.63
Köster 1971	3.23	4.73	5.13
Dravnieks and Laffort 1972	4.24	4.34	4.74
Dravnieks 1974	2.82	4.02	4.42
Hellman and Small 1974	5.10	4.58	4.98
Standard		4.59	4.99

2-Propanone (acetone) (C_3H_6O Mol.Wt.: 58.1 Handbook: p1649)	p.ol mass author	p.ol mass weighted	p.ol vol weighted
Zwaardemaker 1914	5.16	5.08	5.46
Backman 1917	5.38	5.38	5.76
Van Anrooij 1931	5.96	5.22	5.60
Jung 1936	4.11	3.64	4.02
Scherberger *et al.* 1958	2.73	4.15	4.53
Stuiver 1958	5.24	3.95	4.33
Naus 1962	5.40	4.78	5.16
May 1966	3.12	4.31	4.69
Leonardos *et al.* 1969	3.62	3.61	3.99
Dravnieks and Laffort 1972	3.62	3.72	4.10
Hellman and Small 1974	4.32	3.80	4.18
Dravnieks 1974	2.81	4.01	4.39
Takhirov 1974	5.94	5.13	5.51
Makeicheva 1978	6.03	4.93	5.31
Anon 1980	4.15	3.18	3.56
Naus 1982	6.00	5.52	5.90
Punter 1983	5.07	5.47	5.85
Standard		4.46	4.84

propargylaldehyde see: **2-Propynal**

2-Propenal (acrolein) (C_3H_4O Mol.Wt.: 56.1 Handbook: p1707)	p.ol mass author	p.ol mass weighted	p.ol vol weighted
Katz and Talbert 1930	5.39	5.59	5.96
Buchberg *et al.* 1961	6.43	6.55	6.92
Leonardos *et al.* 1969	6.32	6.31	6.68
Knuth 1973	6.85	6.62	6.99
Cormack *et al.* 1974	6.64	6.27	6.64
Teranishi *et al.* 1974	7.30	7.21	7.58
Anon 1980	7.16	6.19	6.56
Standard		6.39	6.76

Propene (propylene) (C_3H_6 Mol.Wt.: 42.1 Handbook: p1713)

	p.ol mass author	p.ol mass weighted	p.ol vol weighted
Laffort and Dravnieks 1973	3.77	4.18	4.42
Hellman and Small 1974	4.42	3.90	4.14
Standard		4.04	4.28

2-Propenenitrile (acrylonitrile) (C_3H_3N Mol.Wt.: 53.1 Handbook: p1776)

	p.ol mass author	p.ol mass weighted	p.ol vol weighted
Stalker 1963	5.47	4.55	4.89
Leonardos *et al.* 1969	4.33	4.32	4.66
Standard		4.43	4.78

2-Propene-l-thiol (allyl mercaptan) (C_3H_6S Mol.Wt.: 74.2 Handbook: p1759)

	p.ol mass author	p.ol mass weighted	p.ol vol weighted
Katz and Talbert 1930	8.05	8.25	8.74
Stuiver 1958	10.85	9.56	10.05
Standard		8.90	9.40

Propenoic acid (acrylic acid) ($C_3H_4O_2$ Mol.Wt.: 72.1 Handbook: p1766)

	p.ol mass author	p.ol mass weighted	p.ol vol weighted
Hellman and Small 1974	6.57	6.05	6.53
Standard		6.05	6.53

2-Propen-l-ol (allyl alcohol) (C_3H_6O Mol.Wt.: 58.1 Handbook: p1814)

	p.ol mass author	p.ol mass weighted	p.ol vol weighted
Katz and Talbert 1930	5.49	5.69	6.07
Jones 1955*b*	4.08	5.59	5.97
Pliska and Janicek 1965	4.32	7.12	7.50
Dravnieks and Laffort 1972	5.92	6.02	6.40
Dravnieks 1974	5.31	6.51	6.89
Standard		6.18	6.57

2-Propenylamine (allylamine) (C_3H_7N Mol.Wt.: 57.1 Handbook: p1714)

	p.ol mass author	p.ol mass weighted	p.ol vol weighted
Katz and Talbert 1930	4.85	5.05	5.43
Standard		5.05	5.43

2-Propenyldithio-2-propene (diallyl disulfide) ($C_6H_{10}S_2$ Mol.Wt.: 146.3 Handbook: d261*)

	p.ol mass author	p.ol mass weighted	p.ol vol weighted
Katz and Talbert 1930	8.14	8.34	9.13
Standard		8.34	9.13

2-Propenyl isocyanide (allylcarbylamine) (C_4H_5N Mol.Wt.: 67.1 Handbook: p1746)

	p.ol mass author	p.ol mass weighted	p.ol vol weighted
Katz and Talbert 1930	7.31	7.51	7.96
Standard		7.51	7.96

2-Propenyl isothiocyanate (allyl isothiocyanate) (C_4H_5NS Mol.Wt.: 99.2 Handbook: i216)

	p.ol mass author	p.ol mass weighted	p.ol vol weighted
Allison and Katz 1919	5.09	7.67	8.29
Katz and Talbert 1930	6.21	6.41	7.03
Stone *et al.* 1967	6.71	6.44	7.06
Standard		6.84	7.46

2-Propenylthio-2-propene (diallyl sulfide) ($C_6H_{10}S$ Mol.Wt.: 114.2 Handbook: s292)

	p.ol mass author	p.ol mass weighted	p.ol vol weighted
Katz and Talbert 1930	9.18	9.38	10.06
Standard		9.38	10.06

propiolaldehyde see: **2-Propynal**

Propyl acetate ($C_5H_{10}O_2$ Mol.Wt.: 102.1 Handbook: a259)

	p.ol mass author	p.ol mass weighted	p.ol vol weighted
Backman 1917	4.92	4.92	5.55
Jung 1936	6.46	5.99	6.62
May 1966	4.15	5.34	5.97
Hellman and Small 1974	6.70	6.18	6.81
Standard		5.61	6.24

Propylamine (C_3H_9N Mol.Wt.: 59.1 Handbook: p1108)

	p.ol mass author	p.ol mass weighted	p.ol vol weighted
Stephens 1971	7.66	7.57	7.96
Standard		7.57	7.96

Propyl butanoate ($C_7H_{14}O_2$ Mol.Wt.: 130.2 Handbook: b2686)

	p.ol mass author	p.ol mass weighted	p.ol vol weighted
Dravnieks and Laffort 1972	6.61	6.71	7.44
Standard		6.71	7.44

Propyldithiopropane (dipropyl disulfide) ($C_6H_{14}S_2$ Mol.Wt.: 150.3 Handbook: d262)

	p.ol mass author	p.ol mass weighted	p.ol vol weighted
Logtenberg 1978	6.88	6.35	7.15
Standard		6.35	7.15

Propylene see: **Propene**
propylene chloride see: **1,2-Dichloropropane**
propylene oxide see: **1,2-Epoxypropane**

Propyl formate ($C_4H_8O_2$ Mol.Wt.: 88.1 Handbook: f146)

	p.ol mass author	p.ol mass weighted	p.ol vol weighted
Backman 1917	4.90	4.90	5.47
Standard		4.90	5.47

Propyl pentanoate ($C_8H_{16}O_2$ Mol.Wt.: 144.2 Handbook: p234)

	p.ol mass author	p.ol mass weighted	p.ol vol weighted
Backman 1917	7.57	7.57	8.35
Standard		7.57	8.35

Propyl propanoate ($C_6H_{12}O_2$ Mol.Wt.: 116.2 Handbook: p1332)

	p.ol mass author	p.ol mass weighted	p.ol vol weighted
Backman 1917	6.61	6.61	7.30
Standard		6.61	7.30

Propylthiopropane (dipropyl sulfide) ($C_6H_{14}S$ Mol.Wt.: 118.2 Handbook: s332)

	p.ol mass author	p.ol mass weighted	p.ol vol weighted
Katz and Talbert 1930	7.28	7.48	8.17
Wilby 1969	6.96	6.88	7.57
Standard		7.18	7.87

2-Propynal (propiolaldehyde,propargyladehyde) (C_3H_2O Mol.Wt.: 54.1 Handbook: p1828)

	p.ol mass author	p.ol mass weighted	p.ol vol weighted
Katz and Talbert 1930	6.46	6.66	7.01
Standard		6.66	7.01

pseudocumene see: **1,2,4-Trimethylbenzene**
pseudocumidine see: **2,4,5-Trimethylaniline**

Pulegone ($C_{10}H_{16}O$ Mol.Wt.: 152.2 Handbook: p1887)

	p.ol mass author	p.ol mass weighted	p.ol vol weighted
Tempelaar 1913	7.08	7.67	8.47
Standard		7.67	8.47

Pyridine (azine) (C_5H_5N Mol.Wt.: 79.1 Handbook: p1939)

	p.ol mass author	p.ol mass weighted	p.ol vol weighted
Hermanides 1909	6.79	6.24	6.76
Zwaardemaker 1914	7.40	7.32	7.84
Backman 1917	6.70	6.70	7.22
Allison and Katz 1919	4.49	7.07	7.59
Katz and Talbert 1930	6.13	6.33	6.85
Van Anrooij 1931	7.11	6.37	6.89
Geier 1936	7.04	6.62	7.14
Jones 1955*b*	4.40	5.91	6.43
Sales 1958	6.37	6.43	6.95
Janicek *et al.* 1960	5.34	6.92	7.44
Leonardos *et al.* 1969	7.17	7.16	7.68
Dravnieks and Laffort 1972	6.48	6.58	7.10
Laffort and Dravnieks 1973	6.13	6.54	7.06
Dravnieks 1974	5.22	6.42	6.94
Laing *et al.* 1978	5.62	6.74	7.26
Naus 1982	6.00	5.52	6.04
Standard		6.56	7.07

Pyrrolidine (azolidine) (C_4H_9N Mol.Wt.: 71.1 Handbook: p2160)

	p.ol mass author	p.ol mass weighted	p.ol vol weighted
Kendall *et al.* 1968	7.44	7.03	7.50
Hendriks 1979	6.00	5.67	6.14
Standard		6.35	6.82

1-Pyrroline (3,4-dihydro-2H-azole) (C_4H_7N Mol.Wt.: 69.1 Handbook: p2173*)

	p.ol mass author	p.ol mass weighted	p.ol vol weighted
Amoore 1977	8.29	7.53	7.99
Hendriks 1979	7.10	6.77	7.23
Standard		7.15	7.61

quinone see: **1,4-Benzoquinone**

Quinoline (C_9H_7N Mol.Wt.: 129.2 Handbook: q34)

	p.ol mass author	p.ol mass weighted	p.ol vol weighted
Geier 1936	7.52	7.10	7.83
Standard		7.10	7.83

safrole see: **1-Allyl-3,4-methylenedioxybenzene**

Salicyl alcohol (α,2-dihydroxytoluene) ($C_7H_8O_2$ Mol.Wt.: 124.2 Handbook: t460)

	p.ol mass author	p.ol mass weighted	p.ol vol weighted
Backman 1917	5.19	5.19	5.90
Standard		5.19	5.90

salicylaldehyde see: **2-Hydroxybenzaldehyde**
skatole see: **3-Methylindole**
styralyl acetate see: **1-Phenylethyl acetate**
styrene see: **Vinylbenzene**
styrene oxide see: **1-Phenyl-1,2-epoxyethane**

Sulfur dichloride (SCl_2 Mol.Wt.: 103.0 Handbook: I, s533)

	p.ol mass author	p.ol mass weighted	p.ol vol weighted
Leonardos *et al.* 1969	8.40	8.39	9.02
Standard		8.39	9.02

Sulfur dioxide (SO_2 Mol.Wt.: 64.1 Handbook: I, s543)

	p.ol mass author	p.ol mass weighted	p.ol vol weighted
Smolczyk and Cobler 1930	5.39	5.60	6.03
Thomas *et al.* 1943	5.84	5.94	6.37
Beck 1959	5.73	5.46	5.89
Shalamberidze 1967	5.79	5.70	6.13
Leonardos *et al.* 1969	5.92	5.91	6.34
Standard		5.73	6.15

α-Terpineol ($C_{10}H_{18}O$ Mol.Wt.: 154.3 Handbook: t49)

	p.ol mass author	p.ol mass weighted	p.ol vol weighted
Hermanides 1909	6.73	6.18	6.99
Zwaardemaker 1914	6.74	6.66	7.47
Backman 1917	6.73	6.73	7.54
Baldus 1936	8.00	6.90	7.71
Standard		6.62	7.43

1,1,2,2-Tetrachloroethane ($C_2H_2Cl_4$ Mol.Wt.: 167.9 Handbook: e267)

	p.ol mass author	p.ol mass weighted	p.ol vol weighted
Lehmann and Schmidt–Kehl 1936	4.70	8.85	6.69
Dravnieks and Laffort 1972	5.79	5.89	6.73
Dravnieks 1974	4.31	5.51	6.35
Standard		5.75	6.59

Tetrachloroethene (perchloroethylene) (C_2Cl_4 Mol.Wt.: 165.8 Handbook: e420)

	p.ol mass author	p.ol mass weighted	p.ol vol weighted
May 1966	3.49	4.68	5.52
Leonardos *et al.* 1969	4.49	4.48	5.32
Anon 1980	4.92	3.95	4.79
Standard		4.37	5.21

Tetrachloromethane (carbon tetrachloride) (CCl_4 Mol.Wt.: 153.8 Handbook: m296)

	p.ol mass author	p.ol mass weighted	p.ol vol weighted
Allison and Katz 1919	2.34	4.92	5.73
Lehmann and Schmidt–Kehl 1936	3.04	4.19	5.00
May 1966	2.90	4.09	4.90
Leonardos *et al.* 1969	3.53	3.52	4.33
Dravnieks and Laffort 1972	3.55	3.65	4.46
Dravnieks 1974	2.43	3.63	4.44
Punter 1983	3.05	3.45	4.26
Standard		3.92	4.73

tetrahydrofuran see: **1,4-Epoxybutane**

Tetrahydrothiophene (thiophane) (C_4H_8S Mol.Wt.: 88.2 Handbook: t224)

	p.ol mass author	p.ol mass weighted	p.ol vol weighted
Deadman and Prigg 1959	8.74	8.42	8.99
Nevers and Oister 1965	8.43	8.43	9.00
Kendall *et al.* 1968	9.35	8.94	9.51
Kniebes *et al.* 1969	8.85	8.65	9.22
Wilby 1969	8.55	8.47	9.04
Laffort and Dravnieks 1973	8.00	8.41	8.98
Blanchard 1976	7.60	8.37	8.94
Moschandreas and Jones 1983	8.74	8.79	9.36
Standard		8.56	9.13

1,2,4,5-Tetramethylbenzene (durene) ($C_{10}H_{14}$ Mol.Wt.: 134.2 Handbook: b943)

	p.ol mass author	p.ol mass weighted	p.ol vol weighted
Backman 1917	7.08	7.08	7.83
Backman 1918	7.06	6.58	7.33
Standard		6.83	7.58

p-thiocresol see: **4-Methylbenzenethiol**
thiophane see: **Tetrahydrothiophene**

Thiophene (C_4H_4S Mol.Wt.: 84.1 Handbook: t187)

	p.ol mass author	p.ol mass weighted	p.ol vol weighted
Backman 1917	5.34	5.34	5.89
Deadman and Prigg 1959	4.67	4.35	4.90
Dravnieks and Laffort 1972	7.52	7.62	8.17
Bedborough and Trott 1979	7.10	6.40	6.95
Standard		5.93	6.48

thiophenol see: **Benzenethiol**

α-Thujone ($C_{10}H_{16}O$ Mol.Wt.: 152.2 Handbook: t259)

	p.ol mass author	p.ol mass weighted	p.ol vol weighted
Köster 1971	4.59	6.09	6.89
Standard		6.09	6.89

thymol see: **2-Isopropyl-5-methylphenol**

Toluene (C_7H_8 Mol.Wt.: 92.2 Handbook: t275)

	p.ol mass author	p.ol mass weighted	p.ol vol weighted
Backman 1917	5.45	5.45	6.03
Backman 1918	5.70	5.22	5.80
Grijns 1919	3.77	5.01	5.59
Schley 1934	5.23	4.86	5.44
Nader 1958	6.41	6.16	6.74
Deadman and Prigg 1959	5.26	4.94	5.52
Naus 1962	5.70	5.08	5.66
Stalker 1963	6.00	5.08	5.66
Gusev 1965	5.66	5.27	5.85
May 1966	3.86	5.05	5.63
Leonardos *et al.* 1969	4.93	4.92	5.50
Dravnieks and O'Donnell 1971	4.35	4.82	5.40
Köster 1971	4.87	6.37	6.95
Dravnieks and Laffort 1972	5.28	5.38	5.96

Toluene *continued*

	p.ol mass author	p.ol mass weighted	p.ol vol weighted
Hellman and Small 1973, 1974	6.23	5.71	6.29
Dravnieks 1974	4.23	5.43	6.01
Anon 1980	5.46	4.49	5.07
Naus 1982	5.70	5.22	5.80
Punter 1983	4.60	5.00	5.58
Standard		5.23	5.81

α-**Toluenethiol** (benzyl mercaptan) (C_7H_8S Mol.Wt.: 124.2 Handbook: t543)

	p.ol mass author	p.ol mass weighted	p.ol vol weighted
Katz and Talbert 1930	7.89	8.09	8.80
Standard		8.09	8.80

m-toluidine see: **3-Methylaniline**

o-toluidine see: **2-Methylaniline**

p-toluidine see: **4-Methylaniline**

4-tolyl acetate see: **4-Methylphenyl acetate**

tolylene diisocyanate see: **2,4-Diisocyanato-1-methylbenzene**

1,2,4-Tribromobenzene ($C_6H_3Br_3$ Mol.Wt.: 314.8 Handbook: b949)

	p.ol mass author	p.ol mass weighted	p.ol vol weighted
Backman 1917	8.31	8.31	9.43
Standard		8.31	9.43

Tribromomethane (bromoform) ($CHBr_3$ Mol.Wt.: 252.8 Handbook: m303)

	p.ol mass author	p.ol mass weighted	p.ol vol weighted
Passy 1893*a*	5.50	5.53	6.55
Backman 1917	5.63	5.63	6.65
Grijns 1919	3.83	5.07	6.09
Rocen 1920	4.53	5.08	6.10
Standard		5.32	6.35

Tributylamine ($C_{12}H_{27}N$ Mol.Wt.: 185.4 Handbook: a920)

	p.ol mass author	p.ol mass weighted	p.ol vol weighted
Laing *et al.* 1978	6.27	7.39	8.28
Standard		7.39	8.28

2,2,2-Trichloroacetic acid ($C_2HCl_3O_2$ Mol.Wt.: 163.4 Handbook: a610)

	p.ol mass author	p.ol mass weighted	p.ol vol weighted
Backman 1917	5.70	5.70	6.53
Standard		5.70	6.53

1,3,5-Trichlorobenzene ($C_6H_3Cl_3$ Mol.Wt.: 181.5 Handbook: b954)

	p.ol mass author	p.ol mass weighted	p.ol vol weighted
Punter 1983	5.21	5.61	6.49
Standard		5.61	6.49

2,2,3-Trichlorobutanal hydrate ($C_4H_7Cl_3O_2$ Mol.Wt.: 192.5 Handbook: b2484)

	p.ol mass author	p.ol mass weighted	p.ol vol weighted
Backman 1917	5.71	5.71	6.61
Standard		5.71	6.61

Trichloroethanal (chloral) (C_2HCl_3O Mol.Wt.: 147.4 Handbook: a52)

	p.ol mass author	p.ol mass weighted	p.ol vol weighted
Leonardos *et al.* 1969	6.55	6.54	7.33
Standard		6.54	7.33

Trichloroethanal hydrate (chloral hydrate) ($C_2H_3Cl_3O_2$ Mol.Wt.: 165.4 Handbook: a56)

	p.ol mass author	p.ol mass weighted	p.ol vol weighted
Backman 1917	7.38	7.38	8.22
Standard		7.38	8.22

1,1,1-Trichloroethane (methylchloroform) ($C_2H_3Cl_3$ Mol.Wt.: 133.4 Handbook: e273)

	p.ol mass author	p.ol mass weighted	p.ol vol weighted
Scherberger *et al.* 1958	2.78	4.20	4.95
May 1966	2.67	3.86	4.61
Kendall *et al.* 1968	4.05	3.64	4.39
Standard		3.90	4.65

Trichloroethene (C_2HCl_3 Mol.Wt.: 131.4 Handbook: e425)

	p.ol mass author	p.ol mass weighted	p.ol vol weighted
Lehmann and Schmidt–Kehl 1936	3.04	4.19	4.93
Scherberger *et al.* 1958	3.39	4.81	5.55
Naus 1962	5.52	4.90	5.64
May 1966	3.36	4.55	5.29
Leonardos *et al.* 1969	3.94	3.93	4.67
Naus 1982	5.52	5.04	5.78
Standard		4.57	5.31

Trichloromethane (chloroform) ($CHCl_3$ Mol.Wt.: 119.4 Handbook: m306)

	p.ol mass author	p.ol mass weighted	p.ol vol weighted
Passy 1893*a*	4.52	4.55	5.25
Tempelaar 1913	2.52	3.11	3.81
Backman 1917	4.83	4.83	5.53
Allison and Katz 1919	2.48	5.06	5.76
Grijns 1919	2.63	3.87	4.57
Rocen 1920	3.13	3.68	4.38
Mitsumoto 1926	3.34	4.38	5.08
Morimura 1934	3.26	4.43	5.13
Schley 1934	4.37	4.00	4.70
Lehman and Schmidt–Kehl 1936	2.91	4.06	4.76
Scherberger *et al.* 1958	2.16	3.58	4.28
Janicek *et al.* 1960	2.43	4.01	4.71
Naus 1962	5.52	4.90	5.60
Dravnieks and Laffort 1972	3.83	3.93	4.63
Dravnieks 1974	2.87	4.07	4.77
Naus 1982	6.30	5.82	6.52
Punter 1983	3.18	3.58	4.28
Standard		4.23	4.93

Trichloromethyl chloroformate (diphosgene) ($C_2Cl_4O_2$ Mol.Wt.: 197.8 Handbook: f167)

	p.ol mass author	p.ol mass weighted	p.ol vol weighted
Schley 1934	8.04	7.67	8.59
Standard		7.67	8.59

2,4,6-Trichlorophenol ($C_6H_3Cl_3O$ Mol.Wt.: 197.5 Handbook: p703)

	p.ol mass author	p.ol mass weighted	p.ol vol weighted
Backman 1917	8.89	8.89	9.81
Kendall *et al.* 1968	7.67	7.26	8.18
Punter 1983	9.79	10.19	11.11
Standard		8.78	9.70

1,1,1-Trichloropropan-2-ol (isopral) ($C_3H_5Cl_3O$ Mol.Wt.: 163.4 Handbook: p1648)

	p.ol mass author	p.ol mass weighted	p.ol vol weighted
Backman 1917	4.76	4.76	5.59
Standard		4.76	5.59

Tridecane ($C_{13}H_{18}$ Mol.Wt.: 184.4 Handbook: t716)

	p.ol mass author	p.ol mass weighted	p.ol vol weighted
Laffort and Dravnieks 1973	4.37	4.78	5.67
Standard		4.78	5.67

Triethylamine ($C_6H_{15}N$ Mol.Wt.: 101.2 Handbook: a925)

	p.ol mass author	p.ol mass weighted	p.ol vol weighted
Tkachev 1970	6.48	5.73	6.36
Hellman and Small 1974	6.39	5.87	6.50
Laing *et al.* 1978	4.92	6.04	6.67
Standard		5.88	6.51

triethyl citrate see: **Triethyl 2-hydroxy-1,2,3-propanetricarboxylate**

Triethyl 2-hydroxy-1,2,3-propanetricarboxylate (triethyl citrate) ($C_{12}H_{20}O_7$ Mol.Wt.: 276.3 Handbook: c432)

	p.ol mass author	p.ol mass weighted	p.ol vol weighted
Backman 1917	5.86	5.86	6.92
Standard		5.86	6.92

Triiodomethane (iodoform) (CHI_3 Mol.Wt.: 393.7 Handbook: m310)

	p.ol mass author	p.ol mass weighted	p.ol vol weighted
Passy 1893*a*	6.68	6.71	7.93
Backman 1917	7.02	7.02	8.24
Allison and Katz 1919	4.74	7.32	8.54
Standard		7.02	8.24

Trimethylamine (C_3H_9N Mol.Wt.: 59.1 Handbook: a939)

	p.ol mass author	p.ol mass weighted	p.ol vol weighted
Tempelaar 1913	5.68	6.27	6.66
Leonardos *et al.* 1969	9.30	9.29	9.68
Stephens 1971	8.86	8.77	9.16
Bedborough and Trott 1979	8.92	8.22	8.61
Anon 1980	9.59	8.62	9.01
Standard		8.23	8.62

N,N,2-Trimethylaniline ($C_9H_{13}N$ Mol.Wt.: 135.2 Handbook: b276*)

	p.ol mass author	p.ol mass weighted	p.ol vol weighted
Backman 1917	6.51	6.51	7.26
Standard		6.51	7.26

N,N,4-Trimethylaniline ($C_9H_{13}N$ Mol.Wt.: 135.2 Handbook: b277*)

	p.ol mass author	p.ol mass weighted	p.ol vol weighted
Backman 1917	6.46	6.46	7.21
Standard		6.46	7.21

2,4,5-Trimethylaniline (pseudocumidine) ($C_9H_{13}N$ Mol.Wt.: 135.2 Handbook: b276)

	p.ol mass author	p.ol mass weighted	p.ol vol weighted
Backman 1917	7.01	7.01	7.76
Huijer 1918	5.10	6.16	6.91
Standard		6.58	7.34

1,2,4-Trimethylbenzene (pseudocumene) (C_9H_{12} Mol.Wt.: 120.2 Handbook: b974)

	p.ol mass author	p.ol mass weighted	p.ol vol weighted
Backman 1917	6.49	6.49	7.19
Backman 1918	6.70	6.22	6.92
Deadman and Prigg 1959	6.15	5.83	6.53
Anon 1980	6.85	5.88	6.58
Standard		6.11	6.81

1,3,5-Trimethylbenzene (mesitylene) (C_9H_{12} Mol.Wt.: 120.2 Handbook: b975)

	p.ol mass author	p.ol mass weighted	p.ol vol weighted
Backman 1917	6.43	6.43	7.13
Deadman and Prigg 1959	6.70	6.38	7.08
Dravnieks and Laffort 1972	5.70	5.80	6.50
Knuth 1973	5.92	5.69	6.39
Dravnieks 1974	4.92	6.12	6.82
Anon 1980	6.74	5.77	6.47
Punter 1983	4.97	5.37	6.07
Standard		5.94	6.64

3,5,5-Trimethyl-2-cyclohexen-1-one (isophorone)
($C_9H_{14}O$ Mol.Wt.: 138.2 Handbook: i180)

	p.ol mass author	p.ol mass weighted	p.ol vol weighted
Hellman and Small 1974	5.96	5.44	6.20
Standard		5.44	6.20

4(2,6,6-Trimethyl-2-cyclohexen-1-yl)-3-buten-2-one (α-ionone)
($C_{13}H_{20}O$ Mol.Wt.: 192.3 Handbook: i95)

	p.ol mass author	p.ol mass weighted	p.ol vol weighted
Zwaardemaker 1914	10.01	9.93	10.83
Henning 1924	10.31	9.50	10.40
Goldenberg 1967	10.53	9.30	10.20
Appell 1969	11.01	8.62	9.52
Standard		9.33	10.24

4(2,6,6-Trimethyl-1-cyclohexen-1-yl)-3-buten-2-one (β-ionone)
($C_{13}H_{20}O$ Mol.Wt.: 192.3 Handbook: i98)

	p.ol mass author	p.ol mass weighted	p.ol vol weighted
Guadagni 1966	10.16	9.92	10.82
Köster 1971	3.94	5.44	6.34
Standard		7.68	8.58

Undecanal ($C_{11}H_{22}O$ Mol.Wt.: 170.3 Handbook: h16)

	p.ol mass author	p.ol mass weighted	p.ol vol weighted
Pliska and Janicek 1965	6.16	8.96	9.81
Randebrock 1971	9.30	7.45	8.30
Teranishi *et al.* 1974	7.46	7.37	8.22
Standard		7.93	8.78

Undecane ($C_{11}H_{24}$ Mol.Wt.: 156.3 Handbook: h19)

	p.ol mass author	p.ol mass weighted	p.ol vol weighted
Mullins 1955	3.43	5.18	5.99
Laffort and Dravnieks 1973	4.64	5.05	5.86
Standard		5.11	5.93

1-Undecanol ($C_{11}H_{24}O$ Mol.Wt.: 172.3 Handbook: h31)

	p.ol mass author	p.ol mass weighted	p.ol vol weighted
Pliska 1962	4.33	6.31	7.17
Standard		6.31	7.17

2-Undecanone ($C_{11}H_{22}O$ Mol.Wt.: 170.3 Handbook: h39)

	p.ol mass author	p.ol mass weighted	p.ol vol weighted
Tempelaar 1913	5.77	6.36	7.21
Backman 1917	7.41	7.41	8.26
Teranishi *et al.* 1974	6.74	6.65	7.50
Standard		6.81	7.66

9-Undecenal (undecylenic aldehyde) ($C_{11}H_{20}O$ Mol.Wt.: 168.3 Handbook: h45*)

	p.ol mass author	p.ol mass weighted	p.ol vol weighted
Randebrock 1971	9.70	7.85	8.70
Standard		7.85	8.70

undecylenic aldehyde see: **9-Undecenal**
valeraldehyde see: **Pentanal**
valeric acid see: **Pentanoic acid**
vanillin see: **4-Hydroxy-3-methoxybenzaldehyde**

Vinyl acetate ($C_4H_6O_2$ Mol.Wt.: 86.1 Handbook: a197)

	p.ol mass author	p.ol mass weighted	p.ol vol weighted
Gofmekler 1960	6.00	5.45	6.00
Hellman and Small 1973, 1974	6.40	5.88	6.43
Standard		5.66	6.22

Vinylbenzene (styrene) (C_8H_8 Mol.Wt.: 104.2 Handbook: s159)

	p.ol mass author	p.ol mass weighted	p.ol vol weighted
Wolf *et al.* 1956	3.98	6.18	6.82
Deadman and Prigg 1959	6.96	6.64	7.28
Stalker 1963	7.13	6.21	6.85
Leonardos *et al.* 1969	6.55	6.54	7.18
Smith and Hochstettler 1969	6.70	6.26	6.90
Dravnieks and Laffort 1972	5.76	5.86	6.50
Hellman and Small 1973, 1974	6.42	5.90	6.54
Dravnieks 1974	5.09	6.29	6.93
Anon 1980	6.85	5.88	6.52
Standard		6.20	6.84

m-xylene see: **1,3-Dimethylbenzene**
o-oxylene see: **1,2-Dimethylbenzene**
p-xylene see: **1,4-Dimethylbenzene**

The manufacturer's authorised representative in the EU for product safety is Oxford University Press España S.A. of El Parque Empresarial San Fernando de Henares, Avenida de Castilla, 2 - 28830 Madrid (www.oup.es/en or product.safety@oup.com). OUP España S.A. also acts as importer into Spain of products made by the manufacturer.
Printed and bound by CPI Group (UK) Ltd, Croydon, CR0 4YY
14/09/2026
02216295-0002